Eclipse Almanac
2031 To 2040

Ten Years of Solar and Lunar Eclipses

COLOR EDITION

ECLIPSIS TERRÆ MAGNA D. 2 AUGUSTI, 1654 È LUNA VISIBILIS.

Fred Espenak

Edition 1.0
October 2020

Eclipse Almanac 2031 To 2040 – Color Edition

Ten Years of Solar and Lunar Eclipses

Astropixels Publishing
P.O. Box 16197
Portal, AZ 85632

www.astropixels.com/pubs

This book may be ordered at: *www.astropixels.com/pubs/EclipseAlmanac.html*

More about solar and lunar eclipses from 2031 to 2040 can be found at EclipseWise:

www.astropixels.com/news/EclipseAlmanac.html

Astropixels Publication Number: AP028

First Edition (Version 1.0a)

ISBN 978-1-941983-28-7

Printed in the United States of America

Front Cover: The first solar eclipse map was published in 1654 by Erhard Weigel of Jena, Germany (courtesy of Michael Zeiler, eclipse-maps.com)

Back Cover: Portrait of Fred Espenak (Copyright ©2018 by Fred Espenak).

Preface

The **Eclipse Almanac** contains maps and diagrams of every solar and lunar eclipse over a ten-year period. This permits the reader to look ahead and easily determine when and where each of these events will be seen. Particular details about each eclipse are included as well as a 25-year table looking further into the future.

Section 1 covers solar eclipses, while Section 2 is devoted to lunar eclipses. Brief explanations are given for the different types of eclipses, and descriptions of the visual appearance of each one is included. Section 3 tabulates the date and time of the Moon's phases over a decade. New Moon and Full Moon phases that coincide with solar and lunar eclipses, respectively, are identified.

The **Eclipse Almanac** series consists of five volumes. Each one covering a single decade as follows:

1) Eclipse Almanac 2031 to 2040   4) Eclipse Almanac 2051 to 2060
2) Eclipse Almanac 2031 to 2040   5) Eclipse Almanac 2061 to 2070
3) Eclipse Almanac 2041 to 2050

Each volume is available in a color edition as well as a more economical black and white edition.

All times listed in the **Eclipse Almanac** are in Universal Time (UT1). This is the modern replacement of Greenwich Mean Time, which was based on mean solar time from Greenwich, England. In comparison, Universal Time is based on Earth's rotation with respect to distant quasars.

For North Americans, the conversion from UT1 to local time is as follows:

Atlantic Standard Time (AST)	= UT1 - 4 hours
Eastern Standard Time (EST)	= UT1 - 5 hours
Central Standard Time (CST)	= UT1 - 6 hours
Mountain Standard Time (MST)	= UT1 - 7 hours
Pacific Standard Time (PST)	= UT1 - 8 hours

If Daylight Saving Time is in effect in the time zone, you must ADD one hour to the standard time.

The conversion of Universal Time to other time zones is easily found on the Internet.

For more information about solar and lunar eclipses from 2021 to 2070, visit *EclipseWise.com* at:

www.astropixels.com/news/EclipseAlmanac.html

Table of Contents

Central Solar Eclipses from 2031 to 2040

The path of every central solar eclipse from 2031 to 2040 is plotted on a world map. The central paths of total eclipses are shaded blue, while annular eclipses are shaded red. For hybrid eclipses, part of the path is shaded blue (total), and part is shaded red (annular). Major cities are plotted as black dots, scaled by population size. (©2020 F. Espenak)

Photo 1–1 The "Diamond Ring Effect" is seen just before totality begins. Total Solar Eclipse of 2018 Jul 03. ©2018 F. Espenak

Section 1: Solar Eclipses

Introduction

The Moon orbits Earth once every 29.5306 days with respect to the Sun. Over the course of its orbit, the Moon's changing position relative to the Sun results in its familiar phases: New Moon > First Quarter > Full Moon > Last Quarter > New Moon. The New Moon phase is the only one not visible because the illuminated side of the Moon then points away from Earth.

The orbit of the Moon is tilted about 5.1° to Earth's orbit around the Sun. The points where the two orbits appear to cross are called the nodes. When the New Moon occurs near one of these nodes, the Moon's shadow falls on a portion of Earth and a solar eclipse is visible from that region.

The Moon's shadow is composed of three cone-shaped components. The outer or penumbral shadow is a zone where the Sun's rays are partially blocked. Nested within the penumbra is the umbral shadow — a region where direct rays from the Sun are completely blocked. The conical umbra tapers to a point beyond which extends an expanding cone called the antumbra. From within this third shadow, the Moon appears smaller than the Sun and is seen in silhouette against the solar disk.

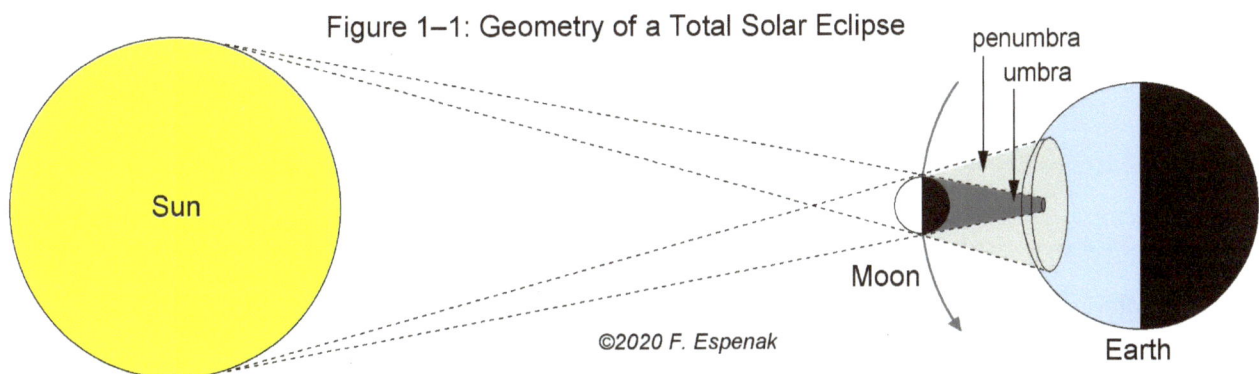

Figure 1–1 illustrates the geometry of a total solar eclipse. A partial eclipse is visible from within the large penumbral shadow, while the total eclipse is only seen from the much smaller umbral shadow.

Photo 1–2 Examples of a partial, annular, and total solar eclipse. ©2020 F. Espenak

Types of Solar Eclipses

There are four types of solar eclipses:

1. **Total Solar Eclipse** — The Moon's penumbral and umbral shadows traverse Earth. The Moon's antumbral shadow extends beyond Earth's surface. The Moon appears larger than the Sun and completely covers the solar disk when viewed from within the umbral shadow. A partial eclipse is seen within the penumbral shadow (Figure 1–1).
2. **Annular Solar Eclipse** — The Moon's penumbral and antumbral shadows traverse Earth. The Moon's umbral shadow completely misses Earth. The Moon's disk appears smaller than the Sun so a bright ring of the Sun's disk surrounds the Moon when viewed from within the antumbral shadow. A partial eclipse is seen within the penumbral shadow (Figure 1–2).
4. **Hybrid Solar Eclipse** — The Moon's penumbral, umbral and antumbral shadows all traverse parts of Earth. The curvature of Earth's surface brings some regions into the umbra and others into the antumbra. The eclipse is total within the umbra and annular within the antumbra. A partial eclipse is seen within the penumbral shadow. Hybrid eclipses are also known as annular-total eclipses.
4. **Partial Solar Eclipse** — The Moon's penumbral shadow traverses Earth while the umbral and antumbral shadows miss Earth. A portion of the Sun's disk is obscured from within the penumbra.

Figure 1–2 illustrates the geometry of an annular solar eclipse. A partial eclipse is visible from within the large penumbral shadow, while the annular eclipse is confined to the much smaller antumbral shadow.

Annular, total and hybrid eclipses are sometimes referred to as *central* eclipses[1]. However, on rare occasions it is possible to have an annular or total eclipse that is non-central. This occurs in Earth's polar regions when only the edge of the umbral or antumbral falls on Earth while the shadow axis misses the planet entirely. The eclipse of 2043 April 09 is a non-central total solar eclipse.

---

[1] A central eclipse is one in which the central axis of the Moon's umbral/antumbral shadow intersects with Earth's surface.

Photo 1–3 Partial solar eclipse of 2014 Oct 23. ©2014 F. Espenak

Visual Appearance of Partial Solar Eclipses

When the Moon's penumbral shadow strikes Earth, a partial eclipse of the Sun is visible from that region. The apparent motion of the Moon with respect to the Sun is gradual — the partial phases can two hours or more. During this time, the Moon's dark limb slowly creeps across the Sun's disk.

Partial eclipses cannot be viewed with the unprotected eye because the Sun is still extremely bright. Special techniques are needed to safely view the eclipse (for example, special eclipse filters or glasses). Even when a partial eclipse reaches its maximum phase, the sky and landscape remain bright. But careful inspection of dappled sunlight beneath a leafy tree will reveal multiple images of the eclipse. The gaps between the tree leaves act like pinhole cameras and project images of the crescent Sun onto the ground.

The Moon's penumbral shadow is typically 4200 to 4500 miles (6700 to 7300 kilometers) in diameter and can cover a significant fraction of the daytime hemisphere of Earth. Consequently, partial eclipses are visible from large geographic regions as the penumbra sweeps across Earth's surface.

Photo 1–4 shows various phases of the Annular solar eclipse of 2005 Oct 03. ©2005 F. Espenak

Visual Appearance of Annular Solar Eclipses

During an annular eclipse, the Moon's penumbral and antumbral shadows sweep across Earth. Compared to the penumbra, the antumbra is much smaller and has a maximum diameter of 232 miles (374 kilometers). Because of this, the antumbra covers a tiny fraction of Earth's surface.

A partial eclipse is visible within the penumbral shadow, but only observers located in the much narrower track of the antumbra will see an annular eclipse. For this reason, the antumbra's trajectory across Earth is called the path of annularity.

All annular eclipses begin with a series of partial phases lasting an hour or more. At the peak of the eclipse, the Moon's disk can be seen in complete silhouette against the Sun. The remaining solar photosphere appears as an intensely bright ring of light surrounding the Moon. The annular phase can last a maximum of 12 ½ minutes but is more typically 3 to 5 minutes in length. After annularity, another series of partial phases occur as the Moon gradually uncovers the Sun.

Special precautions must be used to watch an annular eclipse (just like partial eclipses). Even during the annular phase, the Sun is dangerously bright and cannot be viewed without a solar filter. The landscape and sky remain bright throughout the eclipse, giving little indication of the celestial event in progress.

Photo 1–5 The Sun's corona is only visible to the naked eye for a few minutes during a total eclipse of the Sun. This is an HDR composite image of the total solar eclipse of August 21, 2017 from Casper, Wyoming. © 2017 F. Espenak, www.MrEclipse.com

Visual Appearance of Total Solar Eclipses

During a total eclipse, the Moon's penumbral and umbral shadows fall upon Earth. The umbra has a maximum diameter of 170 miles (273 kilometers). The narrow track traced out as the umbra sweeps across Earth's surface is called the path of totality. The Sun is completely obscured by the Moon from within this zone. The total phase can last up to 7 ½ minutes but is more typically 2 to 3 minutes in length.

Total eclipses all begin and end with a series of partial phases lasting an hour or more. But this is where the resemblance to partial and annular eclipses ends — the total phase is the most spectacular astronomical event visible to the naked eye. During totality, the Sun's outer atmosphere — the solar corona — appears as a gossamer halo surrounding the Moon, and bright stars and planets are visible.

The eclipse takes on a unique character about five minutes before the total phase commences. The Sun has a foreboding quality and casts abnormally sharp shadows. The approaching lunar umbra darkens the western sky and the air temperature is noticeably cooler. A minute before totality, pale shadow bands[2] ripple across the ground.

The ambient light grows feeble even though the crescent Sun is still too bright to see. In the final seconds, the Sun's corona emerges from the glare as the solar crescent shrinks to a brilliant jewel. This celestial diamond ring lingers for a moment before the sunlight is extinguished and totality begins.

The Sun's glorious corona is now displayed to full advantage in the darkened sky. Standing within the Moon's umbra affords the rare and unprecedented opportunity to gaze directly at the glowing million-degree plasma surrounding our star. Twisted, tortured, and constrained by the Sun's enormous magnetic fields, the solar corona is revealed to the naked eye only during the brief seconds when the Moon completely blocks the brilliant disk of the Sun. An eerie twilight bathes the landscape and the colors of dusk surround the horizon.

Minutes race by like seconds. Suddenly, a sparkling bead of sunlight reappears along one edge of the Moon and quickly grows to blindingly bright proportions. Daylight returns as the corona fades and the totality ends. Another hour of partial phases remains before the eclipse is over.

While filters are required for viewing the partial phases, they must be removed for totality. The total phase is the only time it is completely safe to look directly at the Sun without protection. In fact, the total phase is not even visible through solar filters because the Sun's corona is a million times fainter than the photosphere[3].

Figure 1–3: Solar Eclipse Contacts

©2020 F. Espenak

| First Contact (partial eclipse begins) | Second Contact (annular or total eclipse begins) | Third Contact (annular or total eclipse ends) | Fourth Contact (partial eclipse ends) |

Figure 1–3 illustrates the four contacts for annular and total solar eclipses. The arrows indicate the contact point of the Moon with the Sun's limb in each diagram.

Solar Eclipse Contacts

During the course of a solar eclipse, the instants when the edge of the Moon's disk becomes tangent to the Sun's disk are known as eclipse contacts. They mark various stages or phases of a solar eclipse (Figure 1–3).

Partial solar eclipses have two primary contacts.

First Contact (C1) — Partial Eclipse Begins (Instant of first exterior tangency of the Moon with the Sun)
Fourth Contact (C4) — Partial Eclipse Ends (Instant of last exterior tangency of the Moon with the Sun)

---

[2] Shadow bands are caused by the thin solar crescent illuminating Earth's atmosphere before and after totality.
[3] The photosphere is the visible surface of the Sun's disk.

Central solar eclipses (total, annular or hybrid) have four primary contacts. Contacts C2 and C3 mark the instants when the Moon's disk is first and last internally tangent to the Sun. These are the times when the annular or total phase of the eclipse begins and ends, respectively.

First Contact (C1) — Partial Eclipse Begins (Instant of first exterior tangency of the Moon with the Sun)
Second Contact (C2) — Total (or Annular) Eclipse Begins (Instant of first interior tangency of Moon with Sun)
Third Contact (C3) — Total (or Annular) Eclipse Ends (Instant of last interior tangency of Moon with Sun)
Fourth Contact (C4) — Partial Eclipse Ends (Instant of last exterior tangency of the Moon with the Sun)

Photo 1–6 Sequence of partial phases, Diamond Rings, and Totality. Total Solar Eclipse of 2017 Aug 21. ©2017 F. Espenak

Explanation of Global Solar Eclipse Maps

There are 22 eclipses of the Sun during the period 2031 to 2040. A global map is included for each eclipse.

The geographic visibility of each eclipse is illustrated with an orthographic projection map of Earth showing the path of the Moon's penumbral and umbral/antumbral shadows with respect to the continental coastlines, political boundaries (circa 2016), and major cities (represented by black dots). North is up and the daylight terminator is drawn for the instant of greatest eclipse. An asterisk marks the sub-solar point where the Sun appears directly overhead at that time. The salient features of the eclipse maps are identified in the key diagram on page 11.

The limits of the Moon's penumbral shadow delineate the region where a partial solar eclipse is visible. This irregular or saddle shaped region often covers more than half the daylight hemisphere of Earth and consists of several distinct zones or limits. At the northern and/or southern boundaries lie the limits of the penumbra's path. Partial eclipses have only one of these limits, as do central eclipses when the Moon's shadow axis falls no closer than about 0.45 radii from Earth's center. Great loops at the western and eastern extremes of the penumbra's path identify the areas where the eclipse begins/ends at sunrise and sunset, respectively. If the penumbra has both a northern and southern limit, the rising and setting curves form two separate, closed loops (e.g., 2024 Apr 08). Otherwise, the curves are connected in a distorted figure eight (e.g., 2021 Dec 04). Bisecting the *eclipse begins/ends at sunrise and sunset* loops is the curve of maximum eclipse at sunrise (western loop) and sunset (eastern loop).

The eclipse magnitude is defined as the fraction of the Sun's diameter occulted by the Moon. A curve of constant eclipse magnitude delineates the points where the local magnitude at maximum eclipse is equal to a constant value. The maps include *curves of constant eclipse magnitude* for values of 0.2, 0.4, 0.6, and 0.8. These curves run exclusively between the curves of maximum eclipse at sunrise and sunset. They are approximately parallel to the northern/southern penumbral limits and the umbral/antumbral paths of central eclipses. The northern and southern limits of the penumbra may be thought of as curves of eclipse magnitude of 0.0. For total eclipses, the northern and southern limits of the umbra are curves of eclipse magnitude of 1.0.

Greatest eclipse is the instant when the axis of the Moon's shadow cone passes closest to Earth's center. The point on Earth's surface intersected by the axis of the Moon's shadow cone at greatest eclipse is marked by an asterisk. For partial eclipses, the shadow axis misses Earth entirely, so the point of greatest eclipse lies on the day/night terminator and the Sun appears on the horizon.

Parameters for the eclipse appears to the right of each map. They include the time of greatest eclipse (Universal Time[4] or UT1), the eclipse magnitude and obscuration, gamma[5], Saros series, and node (ascending or descending).

Photo 1–7 Time Sequence of the Total Solar Eclipse of 2017 Aug 21. ©2017 F. Espenak

---

[4] Universal Time (UT1) is the modern-day replacement for Greenwich Mean Time.
[5] Gamma is the distance of the Moon's shadow cone axis from Earth's center at greatest eclipse (in Earth equatorial radii).

Global Solar Eclipse Maps

Key to Global Solar Eclipse Maps

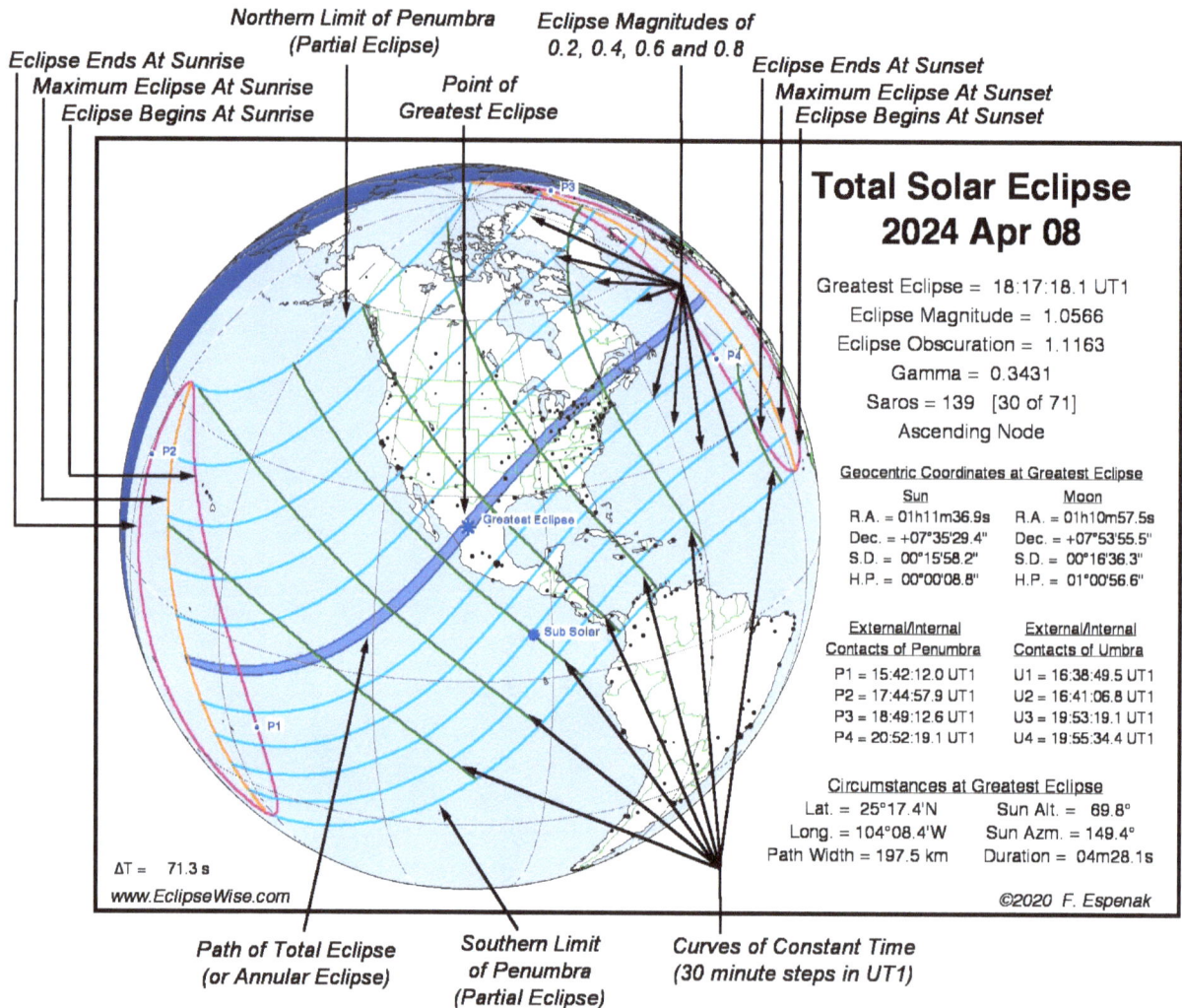

Eclipse Ends At Sunrise
Maximum Eclipse At Sunrise
Eclipse Begins At Sunrise
Northern Limit of Penumbra (Partial Eclipse)
Point of Greatest Eclipse
Eclipse Magnitudes of 0.2, 0.4, 0.6 and 0.8
Eclipse Ends At Sunset
Maximum Eclipse At Sunset
Eclipse Begins At Sunset

Total Solar Eclipse 2024 Apr 08

Greatest Eclipse = 18:17:18.1 UT1
Eclipse Magnitude = 1.0566
Eclipse Obscuration = 1.1163
Gamma = 0.3431
Saros = 139 [30 of 71]
Ascending Node

Geocentric Coordinates at Greatest Eclipse

	Sun	Moon
R.A. =	01h11m36.9s	01h10m57.5s
Dec. =	+07°35'29.4"	+07°53'55.5"
S.D. =	00°15'58.2"	00°16'36.3"
H.P. =	00°00'08.8"	01°00'56.6"

External/Internal Contacts of Penumbra	External/Internal Contacts of Umbra
P1 = 15:42:12.0 UT1	U1 = 16:38:49.5 UT1
P2 = 17:44:57.9 UT1	U2 = 16:41:06.8 UT1
P3 = 18:49:12.6 UT1	U3 = 19:53:19.1 UT1
P4 = 20:52:19.1 UT1	U4 = 19:55:34.4 UT1

Circumstances at Greatest Eclipse

Lat. = 25°17.4'N	Sun Alt. = 69.8°
Long. = 104°08.4'W	Sun Azm. = 149.4°
Path Width = 197.5 km	Duration = 04m28.1s

ΔT = 71.3 s
www.EclipseWise.com
©2020 F. Espenak

Path of Total Eclipse (or Annular Eclipse)
Southern Limit of Penumbra (Partial Eclipse)
Curves of Constant Time (30 minute steps in UT1)

Explanation of Terms Used in Global Solar Eclipse Maps

Greatest Eclipse – The instant when the distance between the axis of the Moon's shadow cone and the center of Earth reaches a minimum (in Universal Time[6] or UT1).

Eclipse Magnitude – The fraction of the Sun's diameter occulted by the Moon at the instant of greatest eclipse (for total and annular eclipses this value is the ratio of diameters of the Moon and the Sun).

Eclipse Obscuration – The fraction of the Sun's area occulted by the Moon at the instant of greatest eclipse.

Gamma – The minimum distance from the lunar shadow axis to the center of Earth (units of Earth equatorial radii).

Saros Series – The Saros series that the eclipse belongs to. The numbers in "[ ]" are the eclipse's sequential position and the number of eclipses in the Saros series.

Node – The orbital node near which the eclipse takes place (Ascending Node or Descending Node).

Geocentric Coordinates of the Sun and the Moon at Greatest Eclipse

> **R.A.** – Right Ascension **S.D.** – Semi-Diameter (i.e. - radius)
> **Dec.** – Declination **H.P.** – Horizontal Parallax

External/Internal Contacts of Penumbra and Umbra – Instants when each shadow enters or exits the surface of Earth (Penumbral contacts shown on map as: P1, P2, P3, and P4; umbral contacts are located at the ends of the central path). All times are in Universal Time (UT1)

Circumstances of Greatest Eclipse – Geographic location (Lat., Long.) of the shadow axis, the altitude and azimuth of the Sun, width of the central path, and the central duration of totality or annularity.

---

[6] Universal Time or UT1 is the modern replacement for Greenwich Mean Time

Annular Solar Eclipse
2031 May 21

Greatest Eclipse = 07:14:49.7 UT1

Eclipse Magnitude = 0.9589

Eclipse Obscuration = 0.9195

Gamma = -0.1970

Saros = 138 [32 of 70]

Decending Node

Geocentric Coordinates at Greatest Eclipse

Sun	Moon
R.A. = 03h51m34.6s	R.A. = 03h51m39.8s
Dec. = +20°09'39.2"	Dec. = +19°58'57.5"
S.D. = 00°15'48.2"	S.D. = 00°14'55.8"
H.P. = 00°00'08.7"	H.P. = 00°54'47.5"

External/Internal Contacts of Penumbra	External/Internal Contacts of Umbra
P1 = 04:14:08.4 UT1	U1 = 05:18:38.0 UT1
P2 = 06:29:22.7 UT1	U2 = 05:22:26.9 UT1
P3 = 08:00:20.1 UT1	U3 = 09:07:13.0 UT1
P4 = 10:15:36.8 UT1	U4 = 09:11:05.9 UT1

Circumstances at Greatest Eclipse

Lat. = 08°55.6'N	Sun Alt. = 78.7°
Long. = 071°43.6'E	Sun Azm. = 353.8°
Path Width = 152.2 km	Duration = 05m25.5s

ΔT = 74.6 s

www.EclipseWise.com

©2020 F. Espenak

Hybrid Solar Eclipse
2031 Nov 14

Greatest Eclipse = 21:06:15.9 UT1

Eclipse Magnitude = 1.0106

Eclipse Obscuration = 1.0213

Gamma = 0.3078

Saros = 143 [24 of 72]

Ascending Node

Geocentric Coordinates at Greatest Eclipse

Sun	Moon
R.A. = 15h19m31.2s	R.A. = 15h19m43.3s
Dec. = -18°20'14.5"	Dec. = -18°02'21.3"
S.D. = 00°16'09.9"	S.D. = 00°16'05.0"
H.P. = 00°00'08.9"	H.P. = 00°59'01.4"

External/Internal Contacts of Penumbra	External/Internal Contacts of Umbra
P1 = 18:23:11.6 UT1	U1 = 19:23:50.9 UT1
P2 = 20:30:56.0 UT1	U2 = 19:24:14.8 UT1
P3 = 21:41:45.2 UT1	U3 = 22:48:22.6 UT1
P4 = 23:49:17.1 UT1	U4 = 22:48:41.5 UT1

Circumstances at Greatest Eclipse

Lat. = 00°37.9'S	Sun Alt. = 72.1°
Long. = 137°38.7'W	Sun Azm. = 188.7°
Path Width = 38.3 km	Duration = 01m08.3s

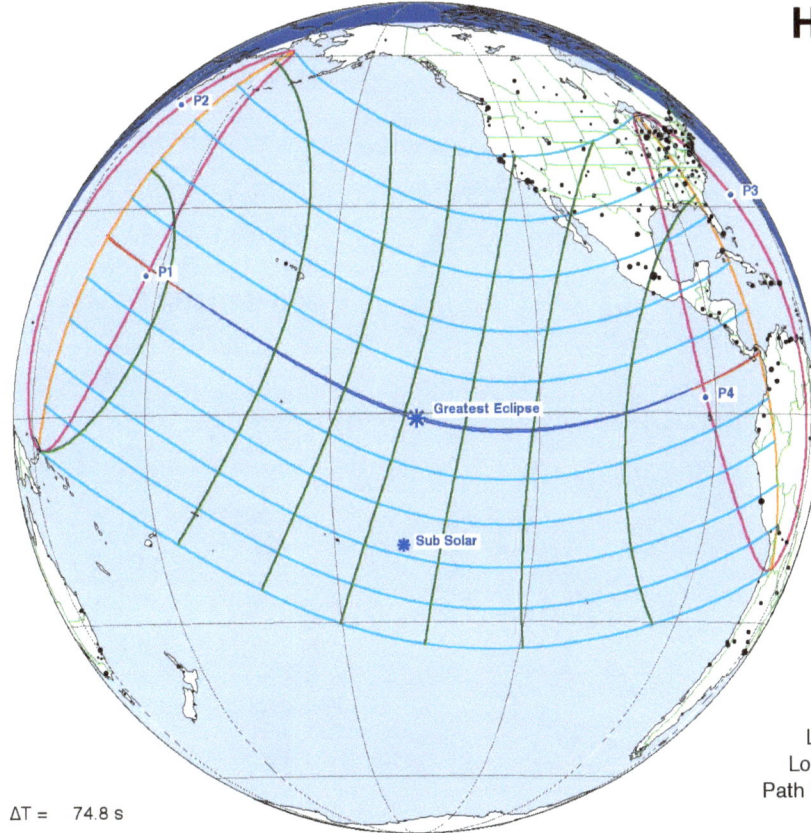

ΔT = 74.8 s

www.EclipseWise.com

©2020 F. Espenak

Annular Solar Eclipse
2032 May 09

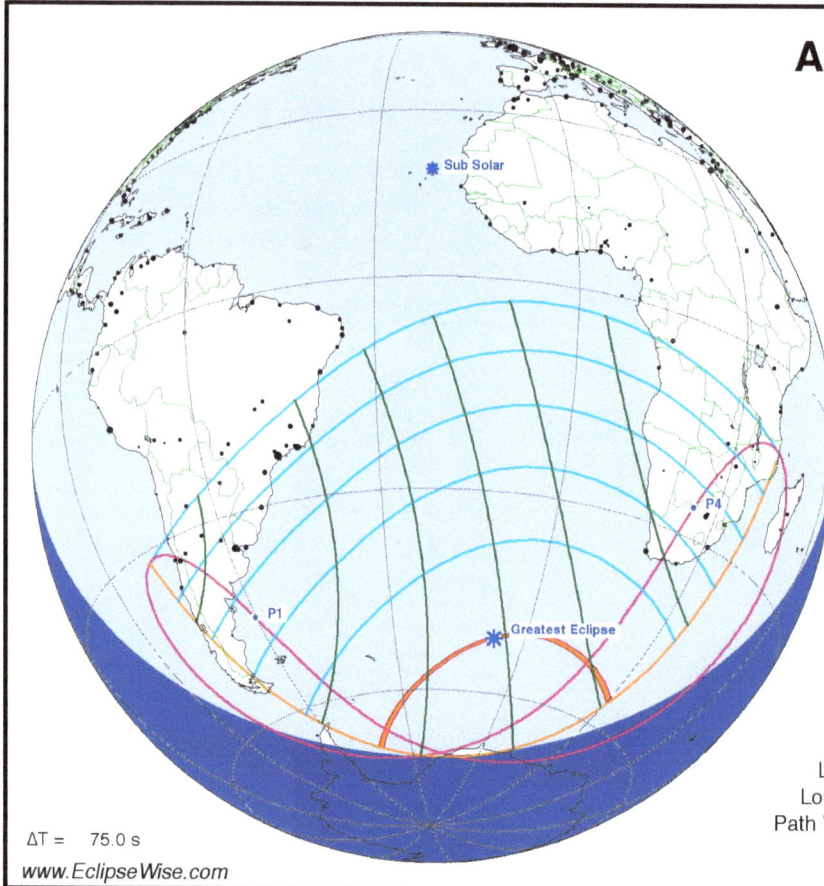

Greatest Eclipse = 13:25:27.3 UT1

Eclipse Magnitude = 0.9957

Eclipse Obscuration = 0.9914

Gamma = -0.9375

Saros = 148 [22 of 75]

Decending Node

Geocentric Coordinates at Greatest Eclipse

	Sun	Moon
R.A. =	03h08m06.7s	R.A. = 03h08m46.1s
Dec. =	+17°35'43.5"	Dec. = +16°42'42.0"
S.D. =	00°15'50.4"	S.D. = 00°15'41.5"
H.P. =	00°00'08.7"	H.P. = 00°57'35.4"

External/Internal Contacts of Penumbra	External/Internal Contacts of Umbra
P1 = 11:09:51.5 UT1	U1 = 12:47:11.5 UT1
P4 = 15:41:17.2 UT1	U2 = 12:48:55.8 UT1
	U3 = 14:02:09.2 UT1
	U4 = 14:03:59.1 UT1

Circumstances at Greatest Eclipse

Lat. = 51°15.6'S	Sun Alt. = 19.9°
Long. = 007°04.4'W	Sun Azm. = 344.6°
Path Width = 43.7 km	Duration = 00m22.0s

ΔT = 75.0 s

www.EclipseWise.com

©2020 F. Espenak

Partial Solar Eclipse
2032 Nov 03

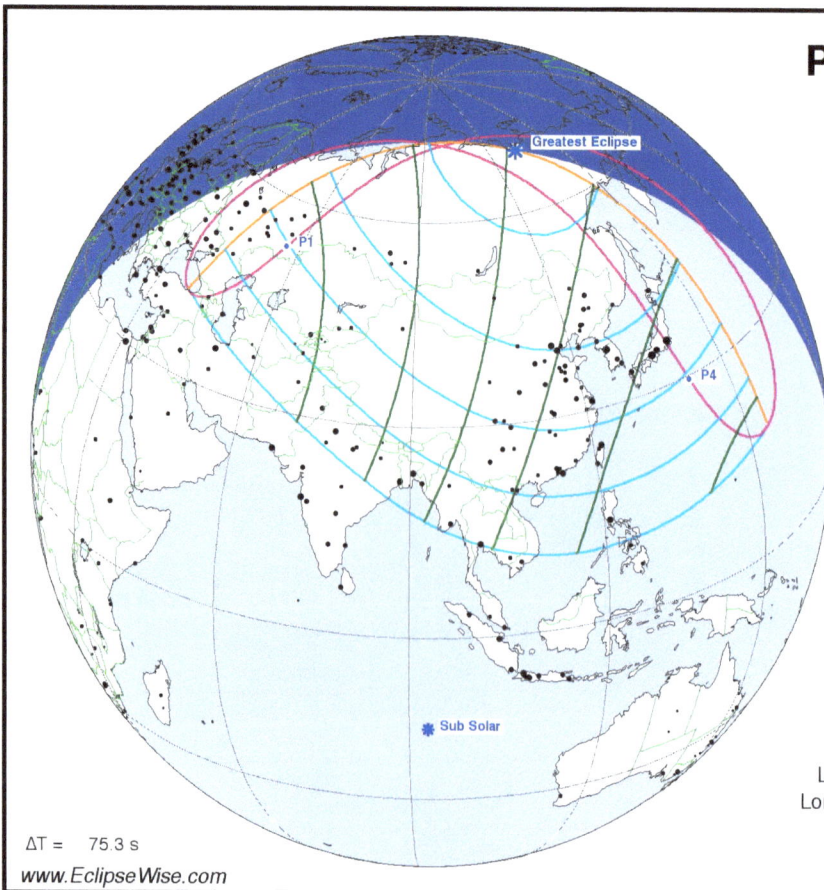

Greatest Eclipse = 05:32:57.6 UT1

Eclipse Magnitude = 0.8554

Eclipse Obscuration = 0.7971

Gamma = 1.0643

Saros = 153 [10 of 70]

Ascending Node

Geocentric Coordinates at Greatest Eclipse

	Sun	Moon
R.A. =	14h35m40.9s	R.A. = 14h36m33.6s
Dec. =	-15°13'54.9"	Dec. = -14°16'01.1"
S.D. =	00°16'07.4"	S.D. = 00°15'13.0"
H.P. =	00°00'08.9"	H.P. = 00°55'50.8"

External/Internal Contacts of Penumbra

P1 = 03:22:20.0 UT1

P4 = 07:43:42.3 UT1

Circumstances at Greatest Eclipse

Lat. = 70°25.7'N	Sun Alt. = 0.0°
Long. = 132°36.5'E	Sun Azm. = 218.3°

ΔT = 75.3 s

www.EclipseWise.com

©2020 F. Espenak

Total Solar Eclipse
2033 Mar 30

Greatest Eclipse = 18:01:20.2 UT1

Eclipse Magnitude = 1.0462

Eclipse Obscuration = 1.0944

Gamma = 0.9778

Saros = 120 [62 of 71]

Decending Node

Geocentric Coordinates at Greatest Eclipse

Sun	Moon
R.A. = 00h38m02.8s	R.A. = 00h36m50.4s
Dec. = +04°05'47.8"	Dec. = +05°02'48.6"
S.D. = 00°16'00.8"	S.D. = 00°16'42.2"
H.P. = 00°00'08.8"	H.P. = 01°01'18.3"

External/Internal Contacts of Penumbra	External/Internal Contacts of Umbra
P1 = 15:59:30.4 UT1	U1 = 17:35:47.2 UT1
P4 = 20:02:55.9 UT1	U2 = 17:48:09.3 UT1
	U3 = 18:14:08.1 UT1
	U4 = 18:26:29.7 UT1

Circumstances at Greatest Eclipse

Lat. = 71°19.0'N	Sun Alt. = 11.2°
Long. = 155°50.1'W	Sun Azm. = 111.1°
Path Width = 781.3 km	Duration = 02m37.0s

ΔT = 75.5 s

www.EclipseWise.com

©2020 F. Espenak

Partial Solar Eclipse
2033 Sep 23

Greatest Eclipse = 13:53:15.5 UT1

Eclipse Magnitude = 0.6890

Eclipse Obscuration = 0.5935

Gamma = -1.1583

Saros = 125 [55 of 73]

Ascending Node

Geocentric Coordinates at Greatest Eclipse

Sun	Moon
R.A. = 12h03m08.9s	R.A. = 12h01m52.5s
Dec. = -00°20'27.7"	Dec. = -01°19'54.7"
S.D. = 00°15'56.3"	S.D. = 00°14'43.6"
H.P. = 00°00'08.8"	H.P. = 00°54'03.0"

External/Internal Contacts of Penumbra
P1 = 11:47:50.8 UT1
P4 = 15:58:21.3 UT1

Circumstances at Greatest Eclipse

Lat. = 72°12.2'S	Sun Alt. = 0.0°
Long. = 121°15.1'W	Sun Azm. = 91.1°

ΔT = 75.7 s

www.EclipseWise.com

©2020 F. Espenak

Total Solar Eclipse
2034 Mar 20

Greatest Eclipse = 10:17:29.2 UT1
Eclipse Magnitude = 1.0458
Eclipse Obscuration = 1.0937
Gamma = 0.2894
Saros = 130 [53 of 73]
Decending Node

Geocentric Coordinates at Greatest Eclipse

	Sun	Moon
R.A. =	23h59m32.7s	R.A. = 23h59m11.3s
Dec. =	-00°02'58.0"	Dec. = +00°13'42.6"
S.D. =	00°16'03.7"	S.D. = 00°16'31.6"
H.P. =	00°00'08.8"	H.P. = 01°00'39.3"

External/Internal Contacts of Penumbra	External/Internal Contacts of Umbra
P1 = 07:39:55.9 UT1	U1 = 08:36:53.6 UT1
P2 = 09:40:05.8 UT1	U2 = 08:38:30.4 UT1
P3 = 10:54:37.9 UT1	U3 = 11:56:18.8 UT1
P4 = 12:54:54.4 UT1	U4 = 11:57:58.9 UT1

Circumstances at Greatest Eclipse

Lat. = 16°03.3'N	Sun Alt. = 73.1°
Long. = 022°13.2'E	Sun Azm. = 161.6°
Path Width = 159.1 km	Duration = 04m09.3s

ΔT = 76.0 s

www.EclipseWise.com

©2020 F. Espenak

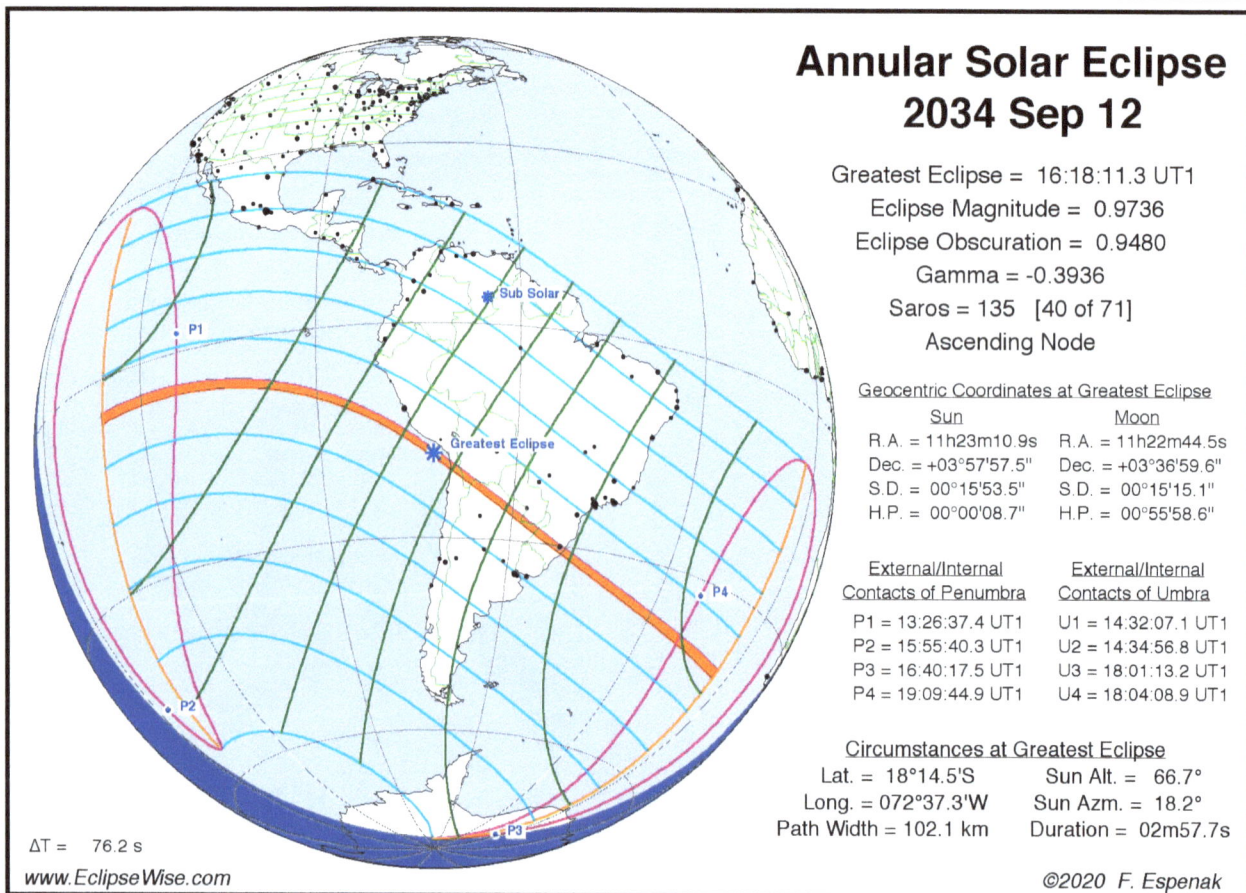

Annular Solar Eclipse
2034 Sep 12

Greatest Eclipse = 16:18:11.3 UT1
Eclipse Magnitude = 0.9736
Eclipse Obscuration = 0.9480
Gamma = -0.3936
Saros = 135 [40 of 71]
Ascending Node

Geocentric Coordinates at Greatest Eclipse

	Sun	Moon
R.A. =	11h23m10.9s	R.A. = 11h22m44.5s
Dec. =	+03°57'57.5"	Dec. = +03°36'59.6"
S.D. =	00°15'53.5"	S.D. = 00°15'15.1"
H.P. =	00°00'08.7"	H.P. = 00°55'58.6"

External/Internal Contacts of Penumbra	External/Internal Contacts of Umbra
P1 = 13:26:37.4 UT1	U1 = 14:32:07.1 UT1
P2 = 15:55:40.3 UT1	U2 = 14:34:56.8 UT1
P3 = 16:40:17.5 UT1	U3 = 18:01:13.2 UT1
P4 = 19:09:44.9 UT1	U4 = 18:04:08.9 UT1

Circumstances at Greatest Eclipse

Lat. = 18°14.5'S	Sun Alt. = 66.7°
Long. = 072°37.3'W	Sun Azm. = 18.2°
Path Width = 102.1 km	Duration = 02m57.7s

ΔT = 76.2 s

www.EclipseWise.com

©2020 F. Espenak

Annular Solar Eclipse
2035 Mar 09

Greatest Eclipse = 23:04:37.1 UT1

Eclipse Magnitude = 0.9919

Eclipse Obscuration = 0.9839

Gamma = -0.4368

Saros = 140 [30 of 71]

Decending Node

Geocentric Coordinates at Greatest Eclipse

Sun	Moon
R.A. = 23h20m17.6s	R.A. = 23h20m47.9s
Dec. = -04°16'22.2"	Dec. = -04°40'23.8"
S.D. = 00°16'06.5"	S.D. = 00°15'44.9"
H.P. = 00°00'08.9"	H.P. = 00°57'47.9"

External/Internal Contacts of Penumbra	External/Internal Contacts of Umbra
P1 = 20:20:47.6 UT1	U1 = 21:25:02.7 UT1
P2 = 22:55:48.6 UT1	U2 = 21:26:39.9 UT1
P3 = 23:13:51.8 UT1	U3 = 00:42:47.6 UT1
P4 = 01:48:27.2 UT1	U4 = 00:44:18.7 UT1

Circumstances at Greatest Eclipse

Lat. = 29°02.6'S	Sun Alt. = 63.9°
Long. = 154°58.0'W	Sun Azm. = 340.2°
Path Width = 31.5 km	Duration = 00m47.5s

ΔT = 76.5 s

www.EclipseWise.com

©2020 F. Espenak

Total Solar Eclipse
2035 Sep 02

Greatest Eclipse = 01:55:29.6 UT1

Eclipse Magnitude = 1.0320

Eclipse Obscuration = 1.0651

Gamma = 0.3727

Saros = 145 [23 of 77]

Ascending Node

Geocentric Coordinates at Greatest Eclipse

Sun	Moon
R.A. = 10h44m07.3s	R.A. = 10h44m32.4s
Dec. = +08°01'09.8"	Dec. = +08°22'14.7"
S.D. = 00°15'50.9"	S.D. = 00°16'06.4"
H.P. = 00°00'08.7"	H.P. = 00°59'06.9"

External/Internal Contacts of Penumbra	External/Internal Contacts of Umbra
P1 = 23:15:29.1 UT1	U1 = 00:15:48.7 UT1
P2 = 01:27:31.9 UT1	U2 = 00:16:50.0 UT1
P3 = 02:23:45.0 UT1	U3 = 03:34:20.9 UT1
P4 = 04:35:41.1 UT1	U4 = 03:35:17.0 UT1

Circumstances at Greatest Eclipse

Lat. = 29°05.6'N	Sun Alt. = 67.9°
Long. = 158°01.0'E	Sun Azm. = 198.5°
Path Width = 116.3 km	Duration = 02m54.2s

ΔT = 76.7 s

www.EclipseWise.com

©2020 F. Espenak

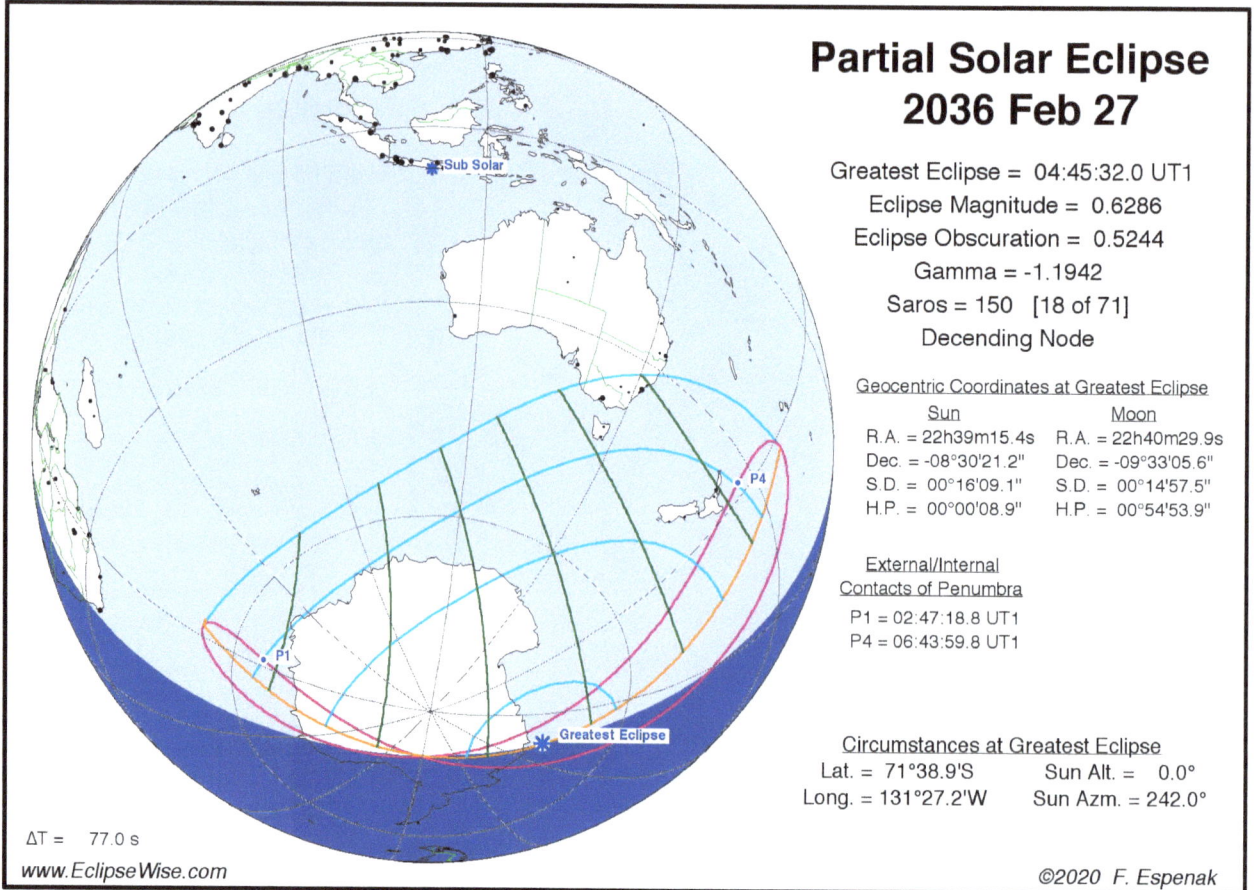

Partial Solar Eclipse
2036 Feb 27

Greatest Eclipse = 04:45:32.0 UT1
Eclipse Magnitude = 0.6286
Eclipse Obscuration = 0.5244
Gamma = -1.1942
Saros = 150 [18 of 71]
Decending Node

Geocentric Coordinates at Greatest Eclipse

	Sun	Moon
R.A. =	22h39m15.4s	R.A. = 22h40m29.9s
Dec. =	-08°30'21.2"	Dec. = -09°33'05.6"
S.D. =	00°16'09.1"	S.D. = 00°14'57.5"
H.P. =	00°00'08.9"	H.P. = 00°54'53.9"

External/Internal
Contacts of Penumbra

P1 = 02:47:18.8 UT1
P4 = 06:43:59.8 UT1

Circumstances at Greatest Eclipse
Lat. = 71°38.9'S   Sun Alt. =   0.0°
Long. = 131°27.2'W   Sun Azm. = 242.0°

ΔT =   77.0 s

www.EclipseWise.com

©2020 F. Espenak

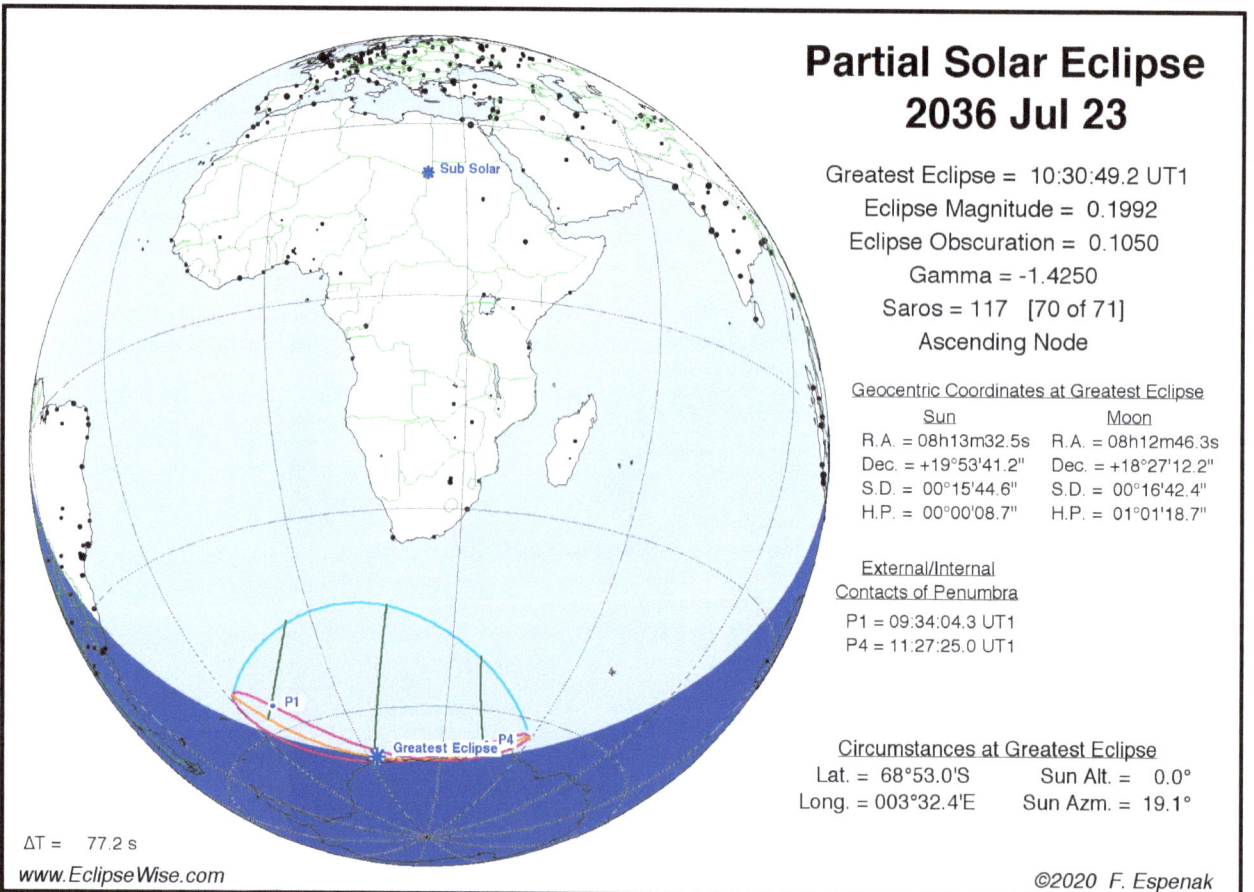

Partial Solar Eclipse
2036 Jul 23

Greatest Eclipse = 10:30:49.2 UT1
Eclipse Magnitude = 0.1992
Eclipse Obscuration = 0.1050
Gamma = -1.4250
Saros = 117 [70 of 71]
Ascending Node

Geocentric Coordinates at Greatest Eclipse

	Sun	Moon
R.A. =	08h13m32.5s	R.A. = 08h12m46.3s
Dec. =	+19°53'41.2"	Dec. = +18°27'12.2"
S.D. =	00°15'44.6"	S.D. = 00°16'42.4"
H.P. =	00°00'08.7"	H.P. = 01°01'18.7"

External/Internal
Contacts of Penumbra

P1 = 09:34:04.3 UT1
P4 = 11:27:25.0 UT1

Circumstances at Greatest Eclipse
Lat. = 68°53.0'S   Sun Alt. =   0.0°
Long. = 003°32.4'E   Sun Azm. = 19.1°

ΔT =   77.2 s

www.EclipseWise.com

©2020 F. Espenak

Partial Solar Eclipse
2036 Aug 21

Greatest Eclipse = 17:24:28.1 UT1

Eclipse Magnitude = 0.8623

Eclipse Obscuration = 0.8381

Gamma = 1.0825

Saros = 155 [7 of 71]

Ascending Node

Geocentric Coordinates at Greatest Eclipse

	Sun	Moon
R.A. =	10h05m24.9s	10h06m34.6s
Dec. =	+11°44'16.4"	+12°48'10.2"
S.D. =	00°15'48.7"	00°16'41.1"
H.P. =	00°00'08.7"	01°01'14.1"

External/Internal Contacts of Penumbra

P1 = 15:33:11.2 UT1
P4 = 19:16:00.3 UT1

Circumstances at Greatest Eclipse

Lat. = 71°07.2'N	Sun Alt. = 0.0°
Long. = 046°59.2'E	Sun Azm. = 308.9°

ΔT = 77.3 s

www.EclipseWise.com

©2020 F. Espenak

Partial Solar Eclipse
2037 Jan 16

Greatest Eclipse = 09:47:37.6 UT1

Eclipse Magnitude = 0.7049

Eclipse Obscuration = 0.6094

Gamma = 1.1477

Saros = 122 [59 of 70]

Decending Node

Geocentric Coordinates at Greatest Eclipse

	Sun	Moon
R.A. =	19h54m30.1s	19h54m05.4s
Dec. =	-20°49'43.5"	-19°47'33.7"
S.D. =	00°16'15.5"	00°14'51.8"
H.P. =	00°00'08.9"	00°54'32.8"

External/Internal Contacts of Penumbra

P1 = 07:41:22.3 UT1
P4 = 11:53:50.9 UT1

Circumstances at Greatest Eclipse

Lat. = 68°31.3'N	Sun Alt. = 0.0°
Long. = 020°47.7'E	Sun Azm. = 166.2°

ΔT = 77.5 s

www.EclipseWise.com

©2020 F. Espenak

19

Total Solar Eclipse
2037 Jul 13

Greatest Eclipse = 02:39:18.1 UT1

Eclipse Magnitude = 1.0413

Eclipse Obscuration = 1.0843

Gamma = -0.7246

Saros = 127 [59 of 82]

Ascending Node

Geocentric Coordinates at Greatest Eclipse

	Sun	Moon
R.A. =	07h31m06.7s	07h30m56.4s
Dec. =	+21°46'57.5"	+21°04'03.1"
S.D. =	00°15'44.0"	00°16'12.0"
H.P. =	00°00'08.7"	00°59'27.3"

External/Internal Contacts of Penumbra

P1 = 00:15:06.5 UT1
P4 = 05:03:23.1 UT1

External/Internal Contacts of Umbra

U1 = 01:25:06.2 UT1
U2 = 01:27:27.1 UT1
U3 = 03:51:03.6 UT1
U4 = 03:53:29.0 UT1

Circumstances at Greatest Eclipse

Lat. = 24°45.8'S   Sun Alt. =   43.4°
Long. = 139°03.3'E   Sun Azm. =   3.3°
Path Width = 200.7 km   Duration = 03m58.4s

ΔT =   77.7 s

www.EclipseWise.com

©2020 F. Espenak

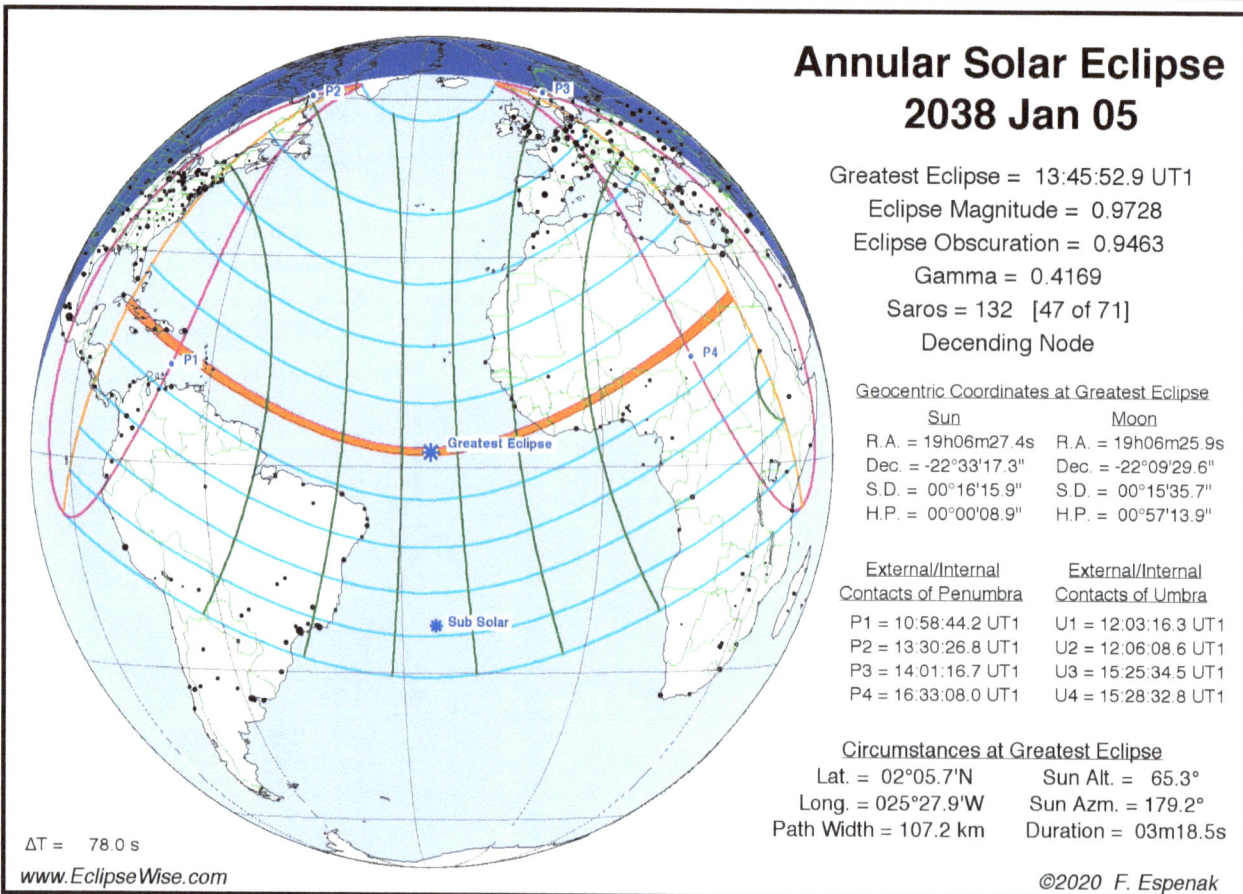

Annular Solar Eclipse
2038 Jan 05

Greatest Eclipse = 13:45:52.9 UT1

Eclipse Magnitude = 0.9728

Eclipse Obscuration = 0.9463

Gamma = 0.4169

Saros = 132 [47 of 71]

Decending Node

Geocentric Coordinates at Greatest Eclipse

	Sun	Moon
R.A. =	19h06m27.4s	19h06m25.9s
Dec. =	-22°33'17.3"	-22°09'29.6"
S.D. =	00°16'15.9"	00°15'35.7"
H.P. =	00°00'08.9"	00°57'13.9"

External/Internal Contacts of Penumbra

P1 = 10:58:44.2 UT1
P2 = 13:30:26.8 UT1
P3 = 14:01:16.7 UT1
P4 = 16:33:08.0 UT1

External/Internal Contacts of Umbra

U1 = 12:03:16.3 UT1
U2 = 12:06:08.6 UT1
U3 = 15:25:34.5 UT1
U4 = 15:28:32.8 UT1

Circumstances at Greatest Eclipse

Lat. = 02°05.7'N   Sun Alt. =   65.3°
Long. = 025°27.9'W   Sun Azm. = 179.2°
Path Width = 107.2 km   Duration = 03m18.5s

ΔT =   78.0 s

www.EclipseWise.com

©2020 F. Espenak

Annular Solar Eclipse
2038 Jul 02

Greatest Eclipse = 13:31:36.8 UT1

Eclipse Magnitude = 0.9911

Eclipse Obscuration = 0.9823

Gamma = 0.0398

Saros = 137 [37 of 70]

Ascending Node

Geocentric Coordinates at Greatest Eclipse

Sun	Moon
R.A. = 06h46m55.4s	R.A. = 06h46m55.2s
Dec. = +22°59'44.2"	Dec. = +23°01'58.2"
S.D. = 00°15'43.9"	S.D. = 00°15'20.9"
H.P. = 00°00'08.6"	H.P. = 00°56'19.9"

External/Internal Contacts of Penumbra	External/Internal Contacts of Umbra
P1 = 10:36:18.3 UT1	U1 = 11:38:00.7 UT1
P2 = 12:41:23.7 UT1	U2 = 11:39:38.2 UT1
P3 = 14:21:50.8 UT1	U3 = 15:23:37.3 UT1
P4 = 16:26:48.9 UT1	U4 = 15:25:09.0 UT1

Circumstances at Greatest Eclipse

Lat. = 25°25.4'N	Sun Alt. = 87.6°
Long. = 021°54.7'W	Sun Azm. = 179.0°
Path Width = 31.2 km	Duration = 00m59.7s

ΔT = 78.3 s

www.EclipseWise.com

©2020 F. Espenak

Total Solar Eclipse
2038 Dec 26

Greatest Eclipse = 00:58:51.1 UT1

Eclipse Magnitude = 1.0269

Eclipse Obscuration = 1.0544

Gamma = -0.2881

Saros = 142 [24 of 72]

Decending Node

Geocentric Coordinates at Greatest Eclipse

Sun	Moon
R.A. = 18h18m51.7s	R.A. = 18h18m46.7s
Dec. = -23°21'47.8"	Dec. = -23°39'05.4"
S.D. = 00°16'15.7"	S.D. = 00°16'25.8"
H.P. = 00°00'08.9"	H.P. = 01°00'18.1"

External/Internal Contacts of Penumbra	External/Internal Contacts of Umbra
P1 = 22:19:32.8 UT1	U1 = 23:17:56.5 UT1
P2 = 00:21:46.3 UT1	U2 = 23:18:33.3 UT1
P3 = 01:35:52.1 UT1	U3 = 02:39:09.9 UT1
P4 = 03:38:12.6 UT1	U4 = 02:39:43.0 UT1

Circumstances at Greatest Eclipse

Lat. = 40°16.7'S	Sun Alt. = 73.0°
Long. = 163°55.7'E	Sun Azm. = 4.5°
Path Width = 95.3 km	Duration = 02m18.2s

ΔT = 78.5 s

www.EclipseWise.com

©2020 F. Espenak

Annular Solar Eclipse
2039 Jun 21

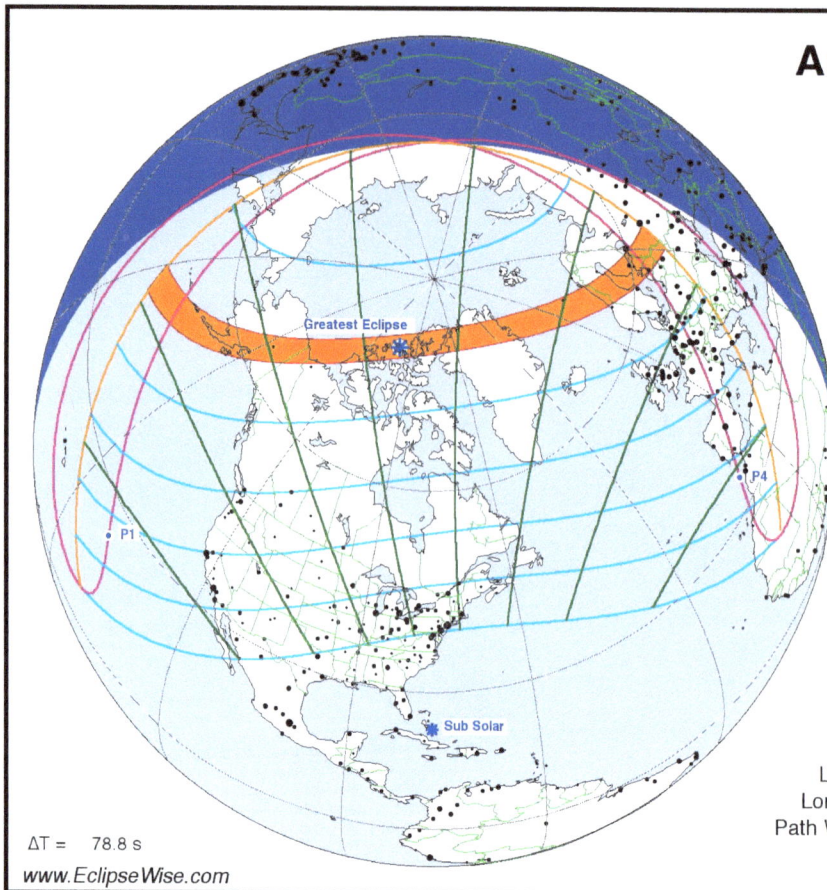

Greatest Eclipse = 17:11:34.9 UT1
Eclipse Magnitude = 0.9454
Eclipse Obscuration = 0.8938
Gamma = 0.8312
Saros = 147 [24 of 80]
Ascending Node

Geocentric Coordinates at Greatest Eclipse

Sun	Moon
R.A. = 06h00m54.5s	R.A. = 06h00m35.3s
Dec. = +23°26'03.6"	Dec. = +24°10'44.9"
S.D. = 00°15'44.3"	S.D. = 00°14'45.6"
H.P. = 00°00'08.7"	H.P. = 00°54'10.2"

External/Internal Contacts of Penumbra	External/Internal Contacts of Umbra
P1 = 14:35:04.9 UT1	U1 = 16:02:25.4 UT1
P4 = 19:47:58.4 UT1	U2 = 16:10:16.8 UT1
	U3 = 18:12:47.6 UT1
	U4 = 18:20:36.8 UT1

Circumstances at Greatest Eclipse

Lat. = 78°52.5'N	Sun Alt. = 33.4°
Long. = 102°07.8'W	Sun Azm. = 152.6°
Path Width = 365.3 km	Duration = 04m04.9s

ΔT = 78.8 s

©2020 F. Espenak

Total Solar Eclipse
2039 Dec 15

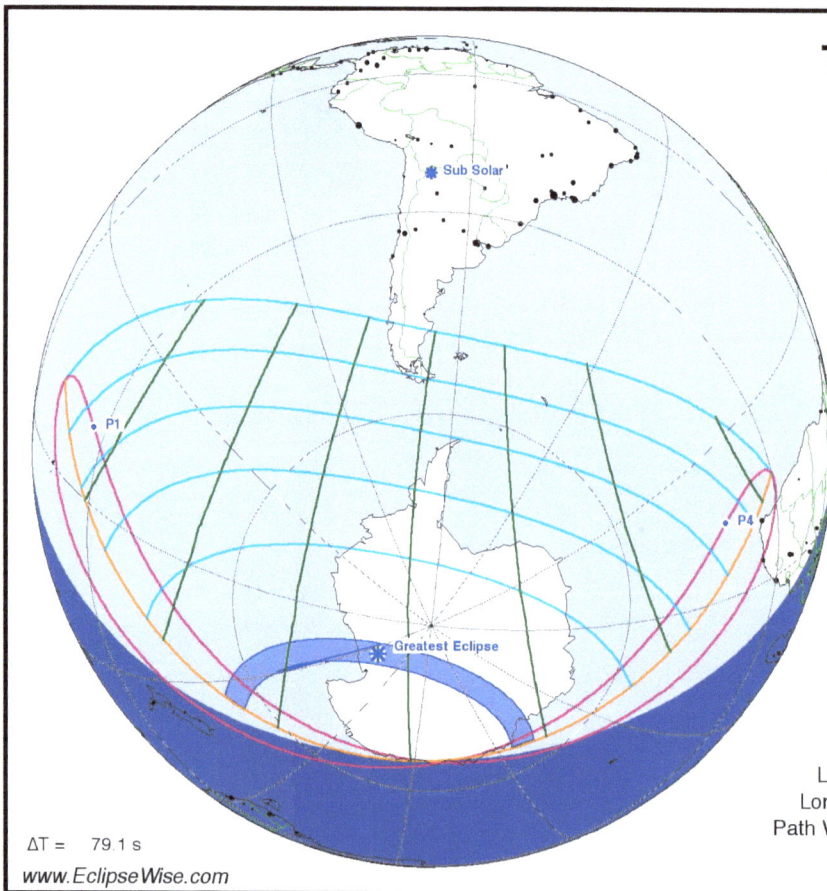

Greatest Eclipse = 16:22:26.8 UT1
Eclipse Magnitude = 1.0356
Eclipse Obscuration = 1.0724
Gamma = -0.9458
Saros = 152 [14 of 70]
Decending Node

Geocentric Coordinates at Greatest Eclipse

Sun	Moon
R.A. = 17h31m51.4s	R.A. = 17h31m14.4s
Dec. = -23°16'37.6"	Dec. = -24°13'58.8"
S.D. = 00°16'14.9"	S.D. = 00°16'44.6"
H.P. = 00°00'08.9"	H.P. = 01°01'26.8"

External/Internal Contacts of Penumbra	External/Internal Contacts of Umbra
P1 = 14:17:38.0 UT1	U1 = 15:47:14.2 UT1
P4 = 18:27:09.0 UT1	U2 = 15:52:23.3 UT1
	U3 = 16:52:20.5 UT1
	U4 = 16:57:30.1 UT1

Circumstances at Greatest Eclipse

Lat. = 80°51.5'S	Sun Alt. = 18.4°
Long. = 172°45.1'E	Sun Azm. = 123.4°
Path Width = 379.8 km	Duration = 01m51.4s

ΔT = 79.1 s

©2020 F. Espenak

Partial Solar Eclipse
2040 May 11

Greatest Eclipse = 03:41:42.8 UT1

Eclipse Magnitude = 0.5306

Eclipse Obscuration = 0.4189

Gamma = -1.2529

Saros = 119 [67 of 71]

Ascending Node

Geocentric Coordinates at Greatest Eclipse

	Sun	Moon
R.A. =	03h14m33.6s	R.A. = 03h16m16.3s
Dec. =	+18°01'19.7"	Dec. = +16°56'30.8"
S.D. =	00°15'50.1"	S.D. = 00°15'06.4"
H.P. =	00°00'08.7"	H.P. = 00°55'26.7"

External/Internal
Contacts of Penumbra

P1 = 01:55:26.0 UT1
P4 = 05:28:26.5 UT1

Circumstances at Greatest Eclipse

Lat. = 62°46.1'S   Sun Alt. =   0.0°
Long. = 174°23.4'E   Sun Azm. = 312.5°

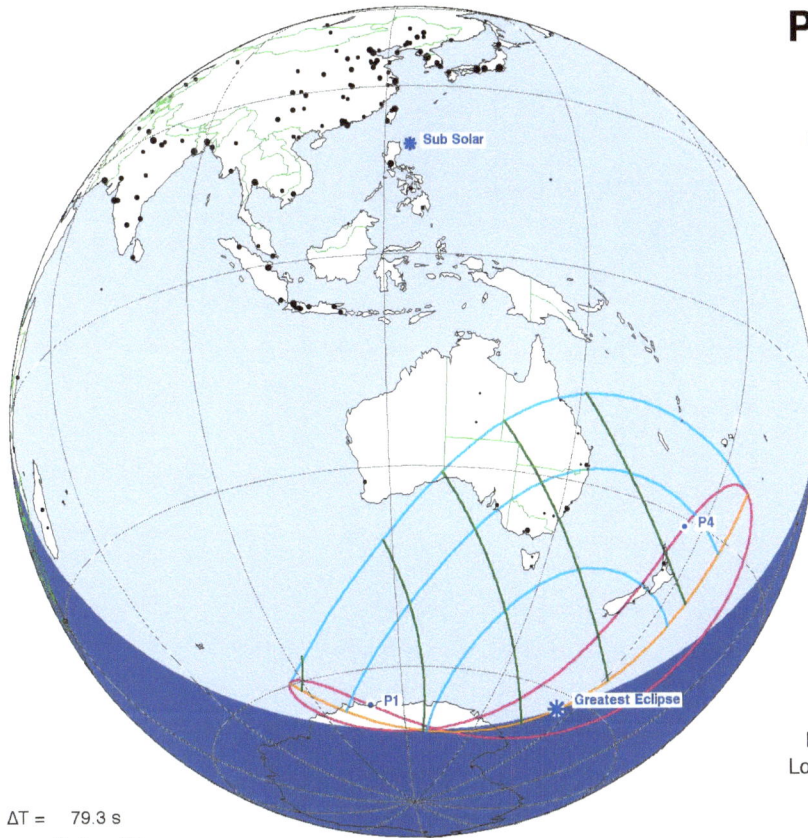

ΔT =   79.3 s

www.EclipseWise.com

©2020 F. Espenak

Partial Solar Eclipse
2040 Nov 04

Greatest Eclipse = 19:07:42.4 UT1

Eclipse Magnitude = 0.8074

Eclipse Obscuration = 0.7513

Gamma = 1.0993

Saros = 124 [56 of 73]

Decending Node

Geocentric Coordinates at Greatest Eclipse

	Sun	Moon
R.A. =	14h42m06.9s	R.A. = 14h43m50.8s
Dec. =	-15°43'53.8"	Dec. = -14°45'19.8"
S.D. =	00°16'07.7"	S.D. = 00°15'49.8"
H.P. =	00°00'08.9"	H.P. = 00°58'05.7"

External/Internal
Contacts of Penumbra

P1 = 17:08:17.9 UT1
P4 = 21:07:22.6 UT1

Circumstances at Greatest Eclipse

Lat. = 62°12.1'N   Sun Alt. =   0.0°
Long. = 053°24.0'W   Sun Azm. = 234.4°

ΔT =   79.6 s

www.EclipseWise.com

©2020 F. Espenak

23

Key to Catalog of Solar Eclipses

The following catalog lists all solar eclipses during a 25-year period.

A brief description of each parameter in the catalog appears below.

Date — Gregorian date of Greatest Eclipse

Greatest Eclipse — Universal Time of Greatest Eclipse

Saros — Saros Series Number of the eclipse

Type — Solar Eclipse Type

      P = Partial Solar Eclipse
      A = Annular Solar Eclipse
      T = Total Solar Eclipse
      H = Hybrid Solar Eclipse

      + = Non-central eclipse of with no northern limit
      – = Non-central eclipse of with no southern limit

      b = Saros series begins (first eclipse in a Saros series)
      e = Saros series ends (last eclipse in a Saros series)

Gamma — minimum distance from the axis of the lunar shadow to the center of Earth

Mag — Eclipse Magnitude; fraction of the Sun's diameter obscured by the Moon

Lat & Long — latitude and longitude where the Sun appears in the zenith at greatest eclipse

Alt — altitude of the Sun at greatest eclipse

Central Line Dur — Central Line Duration (minutes. seconds) at greatest eclipse

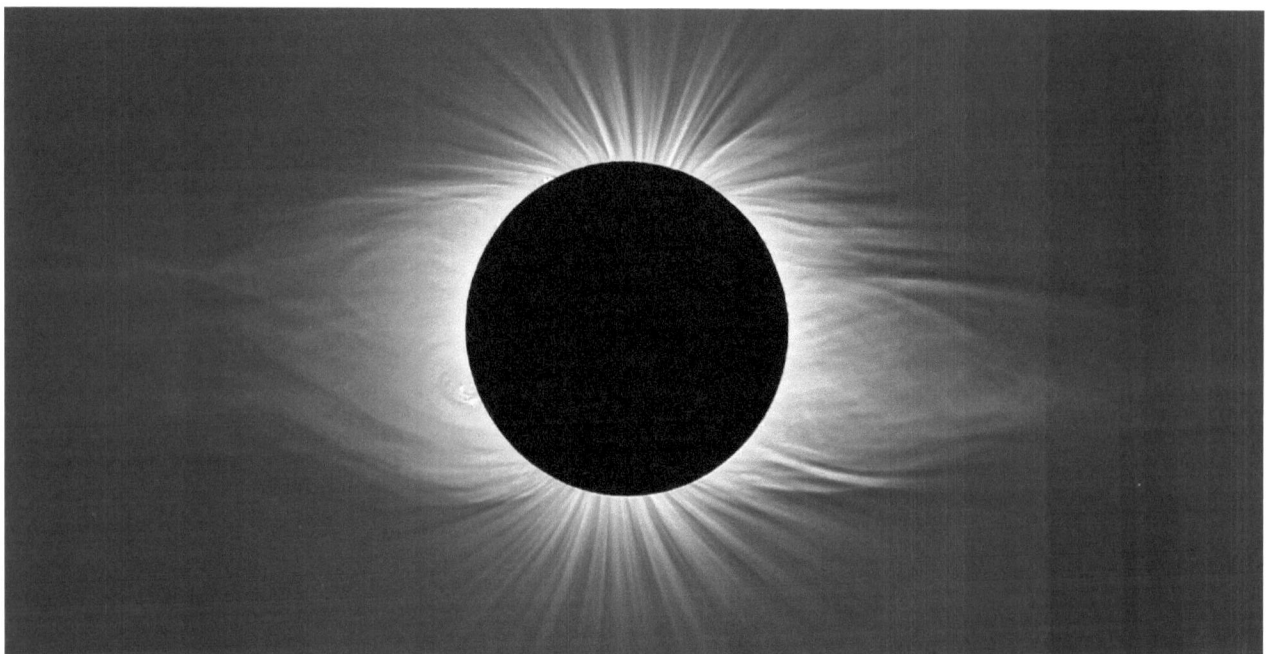

Photo 1–8 The solar corona is revealed during totality. Total Solar Eclipse of 2018 Jul 03. ©2018 F. Espenak

Catalog of Solar Eclipses: 2031 to 2055

Date	Greatest Eclipse	Saros	Type	Gamma	Mag	Lat	Long	Alt	Duration
2031 May 21	7:14:50	138	A	-0.1970	0.9589	8.9N	71.7E	79	05m26s
2031 Nov 14	21:06:16	143	H	0.3078	1.0106	0.6S	137.6W	72	01m08s
2032 May 09	13:25:27	148	A	-0.9375	0.9957	51.3S	7.1W	20	00m22s
2032 Nov 03	5:32:58	153	P	1.0643	0.8554	70.4N	132.6E	0	
2033 Mar 30	18:01:20	120	T	0.9778	1.0462	71.3N	155.8W	11	02m37s
2033 Sep 23	13:53:15	125	P	-1.1583	0.6890	72.2S	121.3W	0	
2034 Mar 20	10:17:29	130	T	0.2894	1.0458	16.1N	22.2E	73	04m09s
2034 Sep 12	16:18:11	135	A	-0.3936	0.9736	18.2S	72.6W	67	02m58s
2035 Mar 09	23:04:37	140	A	-0.4368	0.9919	29.0S	155.0W	64	00m48s
2035 Sep 02	1:55:30	145	T	0.3727	1.0320	29.1N	158.0E	68	02m54s
2036 Feb 27	4:45:32	150	P	-1.1942	0.6286	71.6S	131.5W	0	
2036 Jul 23	10:30:49	117	P	-1.4250	0.1992	68.9S	3.5E	0	
2036 Aug 21	17:24:28	155	P	1.0825	0.8623	71.1N	47.0E	0	
2037 Jan 16	9:47:38	122	P	1.1477	0.7049	68.5N	20.8E	0	
2037 Jul 13	2:39:18	127	T	-0.7246	1.0413	24.8S	139.1E	43	03m58s
2038 Jan 05	13:45:53	132	A	0.4169	0.9728	2.1N	25.5W	65	03m18s
2038 Jul 02	13:31:37	137	A	0.0398	0.9911	25.4N	21.9W	88	01m00s
2038 Dec 26	0:58:51	142	T	-0.2881	1.0269	40.3S	163.9E	73	02m18s
2039 Jun 21	17:11:35	147	A	0.8312	0.9454	78.9N	102.1W	33	04m05s
2039 Dec 15	16:22:27	152	T	-0.9458	1.0356	80.9S	172.8E	18	01m51s
2040 May 11	3:41:43	119	P	-1.2529	0.5306	62.8S	174.4E	0	
2040 Nov 04	19:07:42	124	P	1.0993	0.8074	62.2N	53.4W	0	
2041 Apr 30	11:51:01	129	T	-0.4492	1.0189	9.6S	12.2E	63	01m51s
2041 Oct 25	1:35:02	134	A	0.4133	0.9467	9.9N	162.8E	66	06m07s
2042 Apr 20	2:16:10	139	T	0.2956	1.0614	27.0N	137.3E	73	04m51s
2042 Oct 14	1:59:21	144	A	-0.3030	0.9301	23.7S	137.8E	72	07m44s
2043 Apr 09	18:56:28	149	T+	1.0031	1.0096	61.3N	151.9E	0	01m46s
2043 Oct 03	3:00:28	154	A-	-1.0102	0.9497	61.0S	35.2E	0	02m17s
2044 Feb 28	20:23:18	121	As	-0.9954	0.9600	62.2S	25.6W	4	02m27s
2044 Aug 23	1:15:40	126	T	0.9613	1.0364	64.3N	120.5W	15	02m04s
2045 Feb 16	23:54:45	131	A	-0.3125	0.9285	28.3S	166.2W	72	07m47s
2045 Aug 12	17:41:17	136	T	0.2116	1.0774	25.9N	78.6W	78	06m06s
2046 Feb 05	23:05:03	141	A	0.3765	0.9232	4.8N	171.4W	68	09m42s
2046 Aug 02	10:19:50	146	T	-0.5350	1.0531	12.7S	15.1E	58	04m51s
2047 Jan 26	1:31:55	151	P	1.0450	0.8908	62.9N	111.7E	0	
2047 Jun 23	10:51:07	118	P	1.3766	0.3129	65.8N	178.0W	0	
2047 Jul 22	22:34:54	156	P	-1.3477	0.3605	63.4S	160.1E	0	
2047 Dec 16	23:48:48	123	P	-1.0661	0.8817	66.4S	6.6W	0	
2048 Jun 11	12:57:29	128	A	0.6468	0.9441	63.7N	11.5W	49	04m58s
2048 Dec 05	15:34:02	133	T	-0.3973	1.0440	46.1S	56.4W	66	03m28s
2049 May 31	13:58:34	138	A	-0.1187	0.9631	15.3N	29.9W	83	04m45s
2049 Nov 25	5:32:23	143	H	0.2943	1.0057	3.8S	95.2E	73	00m38s
2050 May 20	20:41:25	148	H	-0.8688	1.0038	40.1S	123.7W	29	00m21s
2050 Nov 14	13:29:27	153	P	1.0447	0.8874	69.5N	1.0E	0	
2051 Apr 11	2:09:13	120	P	1.0169	0.9849	71.6N	32.1E	0	
2051 Oct 04	21:00:48	125	P	-1.2094	0.6024	72.0S	117.7E	0	
2052 Mar 30	18:30:26	130	T	0.3239	1.0466	22.4N	102.6W	71	04m08s
2052 Sep 22	23:37:43	135	A	-0.4480	0.9734	25.7S	174.9E	63	02m51s
2053 Mar 20	7:06:52	140	A	-0.4089	0.9919	23.0S	82.9E	66	00m50s
2053 Sep 12	9:32:41	145	T	0.3140	1.0328	21.5N	41.7E	72	03m04s
2054 Mar 09	12:32:13	150	P	-1.1711	0.6678	72.0S	97.9E	0	
2054 Aug 03	18:02:34	117	Pe	-1.4941	0.0656	69.8S	121.4W	0	
2054 Sep 02	1:08:05	155	P	1.0215	0.9793	71.7N	82.4W	0	
2055 Jan 27	17:52:37	122	P	1.1550	0.6932	69.5N	112.3W	0	
2055 Jul 24	9:56:21	127	T	-0.8012	1.0359	33.3S	25.7E	37	03m17s

Photo 2–1 Phases of the total lunar eclipse of 2014 Apr 15 are captured in this multiple exposure sequence. ©2014 F. Espenak

Section 2: Lunar Eclipses

Introduction

The Moon orbits Earth once every 29.5306 days with respect to the Sun. Over the course of its orbit, the Moon's changing position relative to the Sun results in its familiar phases: New Moon > First Quarter > Full Moon > Last Quarter > New Moon. The New Moon phase is the only one not visible because the illuminated side of the Moon points away from Earth.

During Full Moon, the Moon appears opposite the Sun in the sky. It rises as the Sun sets and is visible throughout the night. The Full Moon sets in the morning just as the Sun rises. This geometry occurs when the Moon is 180° from the Sun as seen from Earth. It corresponds to the direction Earth casts its shadow into space.

The Moon's orbit is tilted about 5.1° to Earth's orbit around the Sun. As seen from Earth, the points where the two orbits appear to cross are called the nodes. When the Full Moon occurs near one of these nodes, the Moon can pass through some portion of Earth's shadow and a lunar eclipse occurs.

Earth's shadow has two cone-shaped components, one nested inside the other. The outer or penumbral shadow is a zone where the Sun's rays are partially blocked., The inner or umbral shadow is a region where direct rays from the Sun are completely blocked.

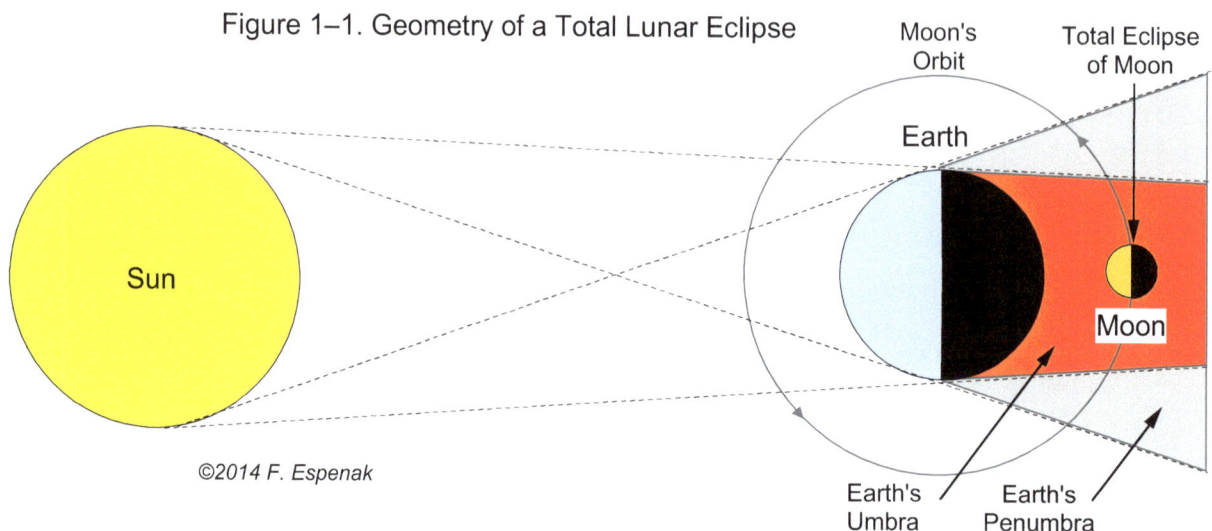

Figure 1–1. Geometry of a Total Lunar Eclipse

Figure 2–1 illustrates the geometry of a total lunar eclipse. A partial eclipse is visible if only part of the Moon enters Earth's umbral shadow. If the Moon passes through the penumbral shadow but misses the umbral shadow, then a penumbral eclipse occurs.

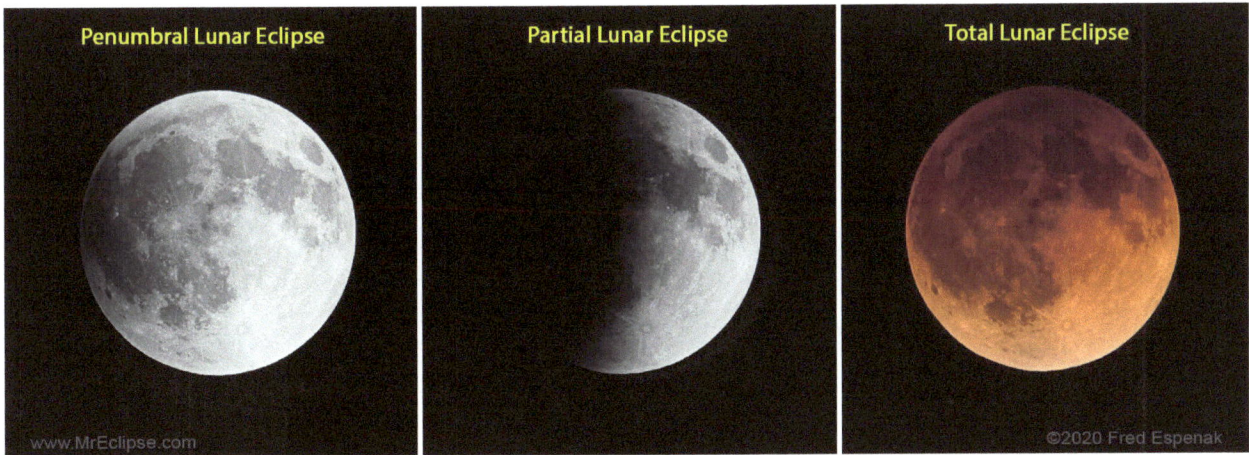

Photo 2–2 Examples of the visual appearance of a penumbral, partial, and total lunar eclipse. ©2020 F. Espenak

Types of Lunar Eclipses

There are three types of lunar eclipses:

1. **Penumbral Lunar Eclipse** — The Moon passes through Earth's faint penumbral shadow. Penumbral eclipses are of minor interest since they are quite subtle and difficult to observe.
2. **Partial Lunar Eclipse** — A portion of the Moon passes through Earth's dark umbral shadow. The remaining part of the Moon appears bright even though it lies deep within the penumbra. Partial eclipses are easy to see, even with the unaided eye.
3. **Total Lunar Eclipse** — The entire Moon passes through Earth's umbral shadow. Total eclipses are quite striking for the vibrant range of color of the Moon during the total phase, referred to as totality.

Figure 1–2. Types of Lunar Eclipses

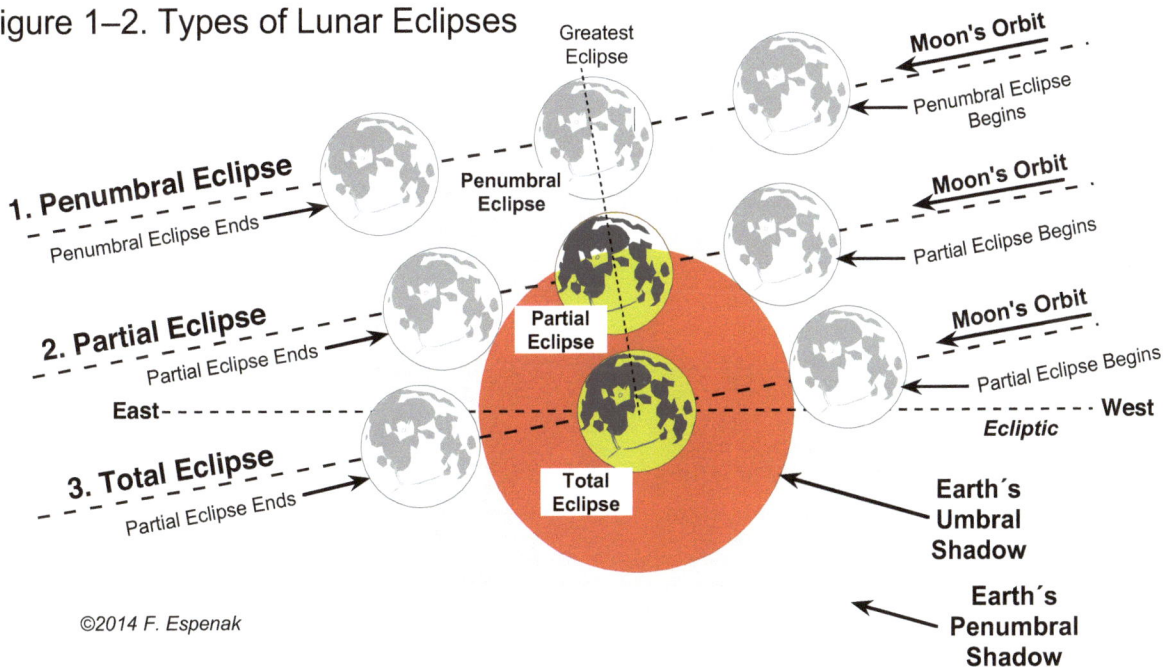

Figure 2–2 illustrates the three types of lunar eclipses as seen from Earth. 1) A penumbral eclipse occurs when the Moon passes through the penumbra but completely misses the umbra. 2) A partial eclipse occurs if some portion of the Moon enters the umbra. 3) A total lunar eclipse takes place when the entire disk of the Moon enters the umbral shadow.

Photo 2–3 Penumbral lunar eclipse of 2002 Nov 20. Left: Moon before the eclipse begins. Right: mid-eclipse when 88.9% of the Moon's is in the penumbral shadow. ©2002 F. Espenak

Visual Appearance of Penumbral Lunar Eclipses

The visual appearance of penumbral, partial and total lunar eclipses differs significantly. While penumbral eclipses are pale and difficult to see, partial eclipses are easy naked-eye events while total eclipses are colorful and dramatic.

Earth's penumbral shadow forms a diverging cone that expands into space away from the Sun. Within this zone, Earth blocks part but not all of the Sun's disk. Some portion of the Sun's direct rays continue to reach the Moon during a penumbral eclipse.

The early and late stages of a penumbral eclipse are completely invisible to the eye. It is only when about 2/3 of the Moon's disk has entered the penumbral shadow that a skilled observer can detect a faint shading across the Moon.

Even when 90% of the Moon is immersed in the penumbra, approximately 10% of the Sun's rays still reach the Moon's deepest limb. Under such conditions, the Moon remains relatively bright with only a subtle shadow gradient across its disk.

Photo 2–4 Time sequence of the partial lunar eclipse of 2012 Jun 04. ©2012 F. Espenak

Visual Appearance of Partial Lunar Eclipses

Compared to penumbral eclipses, partial eclipses are easy to see with the naked eye. The lunar limb extending into the umbral shadow appears very dark or even black. This is due to a contrast effect since the remaining portion of the Moon in the penumbra is hundreds of times brighter. Because the umbral shadow's diameter is about 2.7 times the Moon's diameter, it appears as though a semi-circular bite has been taken out of the Moon.

Aristotle (384–322 BCE) first proved that Earth was round using the curved umbral shadow seen at partial eclipses. In comparing observations of several eclipses, he noted that Earth's shadow was round no matter where the eclipse took place, whether the Moon was high in the sky or low near the horizon. Aristotle reasoned that only a sphere casts a round shadow from every angle.

Photo 2–5 The beginning, middle and end of totality during the total lunar eclipse of 2004 October 28. ©2004 F. Espenak

Visual Appearance of Total Lunar Eclipses

A total lunar eclipse is the most dramatic and visually compelling type of lunar eclipse. The Moon's appearance can vary enormously throughout the period of totality and from one eclipse to the next. The geometry of the Moon's path through the umbra plays a significant role in determining the appearance of totality. The effect that Earth's atmosphere has on a total eclipse is not as apparent. Although the physical mass of Earth blocks all direct sunlight from the umbra, the planet's atmosphere filters, attenuates and bends some of the Sun's rays into the shadow.

The molecules in Earth's atmosphere scatter short wavelength light (i.e., yellow, green, blue) more than long wavelength light (i.e., orange, red). The same process responsible for making sunsets red also gives total lunar eclipses their characteristic ruddy color. The exact appearance can vary widely in both hue and brightness.

Because the lowest layers of the atmosphere are the thickest, they absorb more sunlight and refract it through larger angles. About 75% of the atmosphere's mass is concentrated in the bottom 10 kilometers (troposphere) as well as most of the water vapor, which can form massive clouds that block even more light. Just above the troposphere lies the stratosphere (10 to 50 kilometers), a rarified zone above most of the planet's weather systems. The stratosphere is subject to important photochemical reactions due to the high level of solar ultraviolet radiation that penetrates the region. The troposphere and stratosphere act together as a ring-shaped lens that refracts heavily reddened sunlight into Earth's umbral shadow. Since the higher stratospheric layers contain less gas, they refract sunlight through progressively smaller angles into the outer parts of the umbra. In contrast, denser tropospheric layers refract sunlight through larger angles to reach the inner parts of the umbra.

Because of this lensing effect, the amount of light refracted into the umbra tends to decrease radially from the edge to the center. Inhomogeneities from varying amounts of cloud and dust at differing latitudes can cause significant variations in brightness throughout the umbra.

Besides water (cloud, mist, precipitation), Earth's atmosphere also contains aerosols or tiny particles of organic debris, meteoric dust, volcanic ash and photochemical droplets. This material attenuates sunlight before it is refracted into the umbra. For instance, major volcanic eruptions in 1963 (Agung) and 1982 (El Chichon) each dumped large quantities of gas and ash into the stratosphere and were followed by several years of dark eclipses.

The 1991 eruption of Pinatubo in the Philippines had a similar effect. While most of the solid ash fell to Earth several days after circulating through the troposphere, a sizable volume of sulfur dioxide (SO_2) reached the stratosphere where it interacts with water vapor and eventually produces sulfuric acid (H_2SO_4). This high-altitude volcanic haze layer severely attenuates sunlight that must travel several hundred kilometers horizontally through the layer before being refracted into the umbral shadow. Thus, total eclipses following large volcanic eruptions are particularly dark. The total lunar eclipse of 1992 Dec 09 (1½ years after Pinatubo) was so dark that it was difficult to see the Moon's dull gray disk with the naked eye.

All total eclipses begin with penumbral and partial phases. After the total phase, the eclipse ends with more partial and penumbral phases. Lunar eclipses are completely safe to view and require none of the precautions needed for viewing solar eclipses (like special filters). The best views of a lunar eclipse are with binoculars and the naked eye.

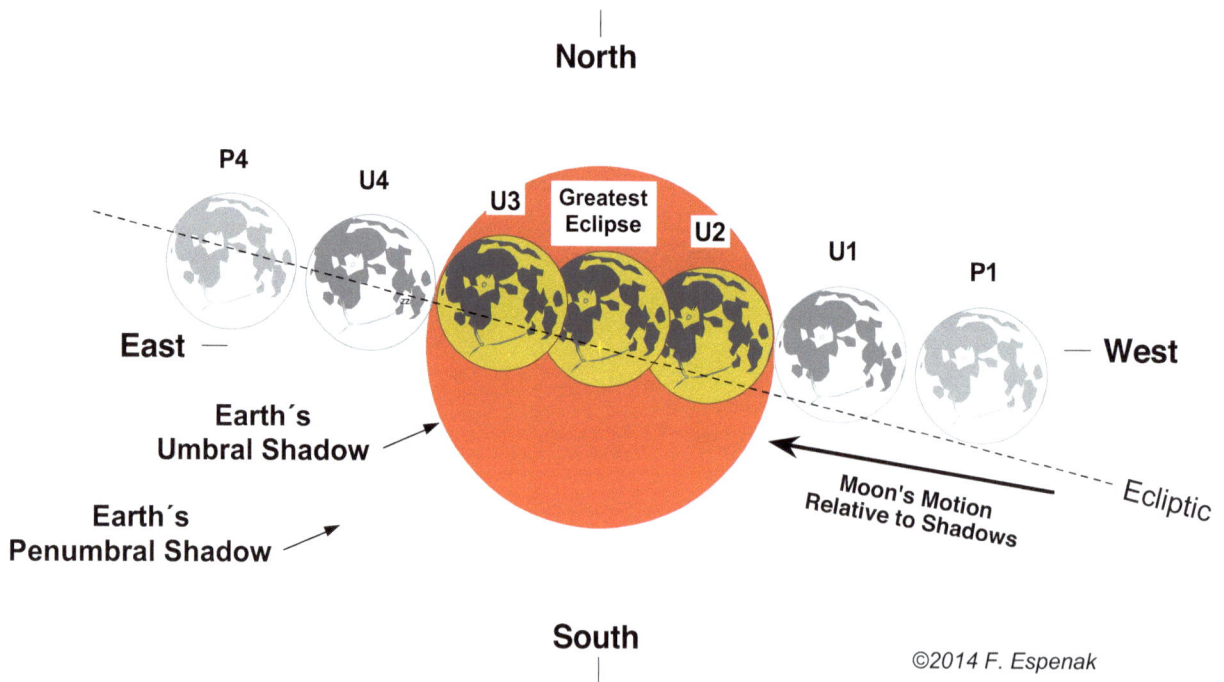

Figure 2–1. Lunar Eclipse Contacts

©2014 F. Espenak

Figure 2–3 illustrates the six contacts for a total lunar eclipse. These correspond to the instants when the Moon's disk is externally tangent to the penumbra (P1 and P4), or either externally or internally tangent to the umbra (U1, U2, U3, and U4). Partial eclipses do not have contacts U2 and U3, while penumbral eclipses only have contacts P1 and P4.

Lunar Eclipse Contacts

During the course of a lunar eclipse, the instants when the Moon's disk becomes tangent to Earth's shadows are known as eclipse contacts. They mark the primary stages or phases of a lunar eclipse (see figure 2-3) and are defined as follows.

P1 — Penumbral Eclipse Begins	(Instant of first exterior tangency of the Moon with the Penumbra)
U1 — Partial Eclipse Begins	(Instant of first exterior tangency of the Moon with the Umbra)
U2 — Total Eclipse Begins	(Instant of first interior tangency of the Moon with the Umbra)
U3 — Total Eclipse Ends	(Instant of last interior tangency of the Moon with the Umbra)
U4 — Partial Eclipse Ends	(Instant of last exterior tangency of the Moon with the Umbra)
P4 — Penumbral Eclipse Ends	(Instant of last exterior tangency of the Moon with the Penumbra)

Penumbral eclipses only have first and last contacts (i.e., P1 and P4) although neither of these events is observable.

In addition to P1 and P4, partial eclipses also have contacts U1 and U4 when the partial phases begin and end.

Total lunar eclipses have all six contacts. Contacts U2 and U3 mark the instants when the Moon's entire disk is first and last internally tangent to the umbra. These are the times when the total phase of the eclipse begins and ends..

The instant of greatest eclipse occurs when the Moon passes closest to the shadow axis. This is the maximum phase of the eclipse when the Moon is at its deepest position within either the penumbral or umbral shadow.

Photo 2–6 Five minutes before the start of totality (2018 Jan 31), the Moon is bathed in an orange-red light.
The narrow rim outside the umbra and still in sunlight appears brilliant white. ©2018 F. Espenak

Enlargement of Earth's Shadows

In 1707, Philippe de La Hire made a curious observation about Earth's umbra. The predicted radius of the shadow needed to be enlarged by about 1/41 in order to fit timings made during a lunar eclipse. The enlargement is attributed to Earth's atmosphere, which becomes increasingly less transparent at lower levels. Additional observations over the next two centuries revealed that the shadow enlargement was somewhat variable from one eclipse to the next.

William Chauvenet (1891) formulated one method to account for the shadow enlargement while André-Louis Danjon (1951) devised another. Both methods assumed a circular cross section for the umbral shadow. However, Earth is flattened at the poles and bulges at the Equator, so an oblate spheroid more closely represents its shape. The projection of each of the planet's shadows is an ellipse rather than a circle. Furthermore, Earth's axial tilt towards or away from the Sun throughout the year means the elliptical shape of the penumbral and umbral shadows varies as well.

In an analysis of 22,539 observations made at 94 lunar eclipses from 1842 to 2011, Herald and Sinnott[7] (2014) found that the size and shape of the umbra are consistent with an oblate spheroid at the time of each eclipse, enlarged by the empirically determined occulting layer that uniformly surrounds Earth. The effective height of this layer was found to be 87 kilometers. Based on this work, the authors developed a new method to calculate the shadow enlargement including their elliptical shape.

This new method is the most rigorous and accurate procedure to date. The *Eclipse Almanac* uses it in the lunar eclipse predictions presented here.

Explanation of Lunar Eclipse Figures

There are 23 eclipses of the Moon during the period 2031 to 2040. A figure for each eclipse is included.

Each figure consists of two diagrams. The first one depicts the Moon's path through Earth's penumbral and umbral shadows (with Celestial North up). The second is a map showing the geographic visibility of each eclipse phase. All features in these diagrams are identified in the key on the next page.

The Moon's orbital motion with respect to the shadows is from west to east (right to left). Each phase of the eclipse is defined by the instant when the Moon's limb is externally or internally tangent to the penumbra or umbra. The six primary contacts of the Moon with the penumbral and umbral shadows are defined as follows.

P1 — Penumbral Eclipse Begins   (Instant of first exterior tangency of the Moon with the Penumbra)
U1 — Partial Eclipse Begins   (Instant of first exterior tangency of the Moon with the Umbra)
U2 — Total Eclipse Begins   (Instant of first interior tangency of the Moon with the Umbra)
U3 — Total Eclipse Ends   (Instant of last interior tangency of the Moon with the Umbra)
U4 — Partial Eclipse Ends   (Instant of last exterior tangency of the Moon with the Umbra)
P4 — Penumbral Eclipse Ends   (Instant of last exterior tangency of the Moon with the Penumbra)

Penumbral lunar eclipses have two primary contacts: P1 and P4, but neither is observable because the edge of the penumbra is indistinct and extremely faint.

In addition to the penumbral contacts, partial lunar eclipses have two more contacts as the Moon's limb enters and exits the umbral shadow: U1 and U4, respectively. These two contacts mark the instants when the partial phase of the eclipse begins and ends.

Total lunar eclipses undergo all six contacts. The two additional umbral contacts are the instants when the Moon's entire disk is first and last internally tangent to the umbra: U2 and U3, respectively. They mark the times when the total phase of the eclipse begins and ends.

The Moon passes closest to the shadow axis at the instant of greatest eclipse. This corresponds to the maximum phase of the eclipse and the Moon's position at this instant is also depicted in the path diagrams.

The equidistant cylindrical projection map shows the geographic region of visibility at each phase of the eclipse. This is accomplished using a series of curves showing where Moonrise and Moonset occur at each eclipse contact. The map is shaded to indicate eclipse visibility. The entire eclipse is visible from the zone with no shading. Conversely, none of the eclipse can be seen from the zone with the darkest shading.

At greatest eclipse, the Moon is deepest in Earth's shadow. The geographic location where the Moon appears in the zenith at greatest eclipse is shown by an asterisk.

Parameters relevant to the eclipse appear in on the right side of each figure. The instant of greatest eclipse is the time (Universal Time[8] or UT1) when the Moon passes closest to the shadow axis. The penumbral and umbral magnitudes are the fractions of the Moon's diameter immersed in each shadow. Gamma is the minimum distance of the Moon's center from the axis of Earth's shadow at greatest eclipse. The Saros series of the eclipse, and the

---

[7] Herald, D., and Sinnott, R. W., "Analysis of Lunar Crater Timings, 1842–2011," *J. Br. Astron. Assoc.*, **124**, 5 (2014)

[8] Universal Time (UT1) is the modern-day replacement for Greenwich Mean Time and is based on Earth's rotation with respect to distant quasars.

node of the Moon's orbit are given. Each contact time of the Moon's edge with the penumbral and umbral shadows is listed in Universal Time (UT1). Depending on the eclipse type, the duration of the penumbral, partial or total phases are given.

2–7 A time sequence shows the partial phases and totality during the total lunar eclipse of 2000 July 16 from Maui, Hawaii (Copyright ©2000 by Fred Espenak).

Lunar Eclipse Figures

Key to Lunar Eclipse Figures

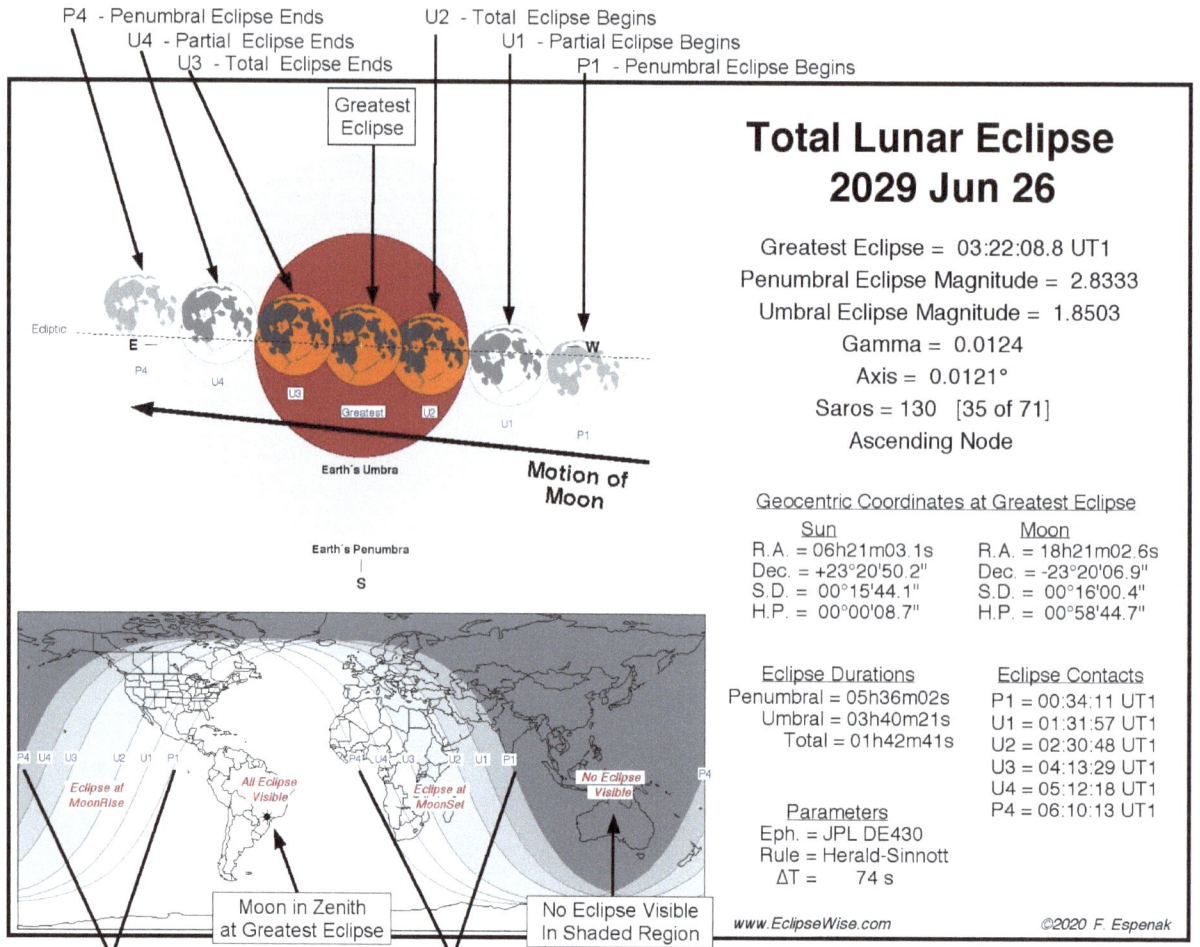

P4 - Penumbral Eclipse Ends
U4 - Partial Eclipse Ends
U3 - Total Eclipse Ends

U2 - Total Eclipse Begins
U1 - Partial Eclipse Begins
P1 - Penumbral Eclipse Begins

Greatest Eclipse

Ecliptic
E
P4
U4
U3
Greatest
U2
U1
P1
W

Earth's Umbra

Motion of Moon

Earth's Penumbra

S

Total Lunar Eclipse
2029 Jun 26

Greatest Eclipse = 03:22:08.8 UT1
Penumbral Eclipse Magnitude = 2.8333
Umbral Eclipse Magnitude = 1.8503
Gamma = 0.0124
Axis = 0.0121°
Saros = 130 [35 of 71]
Ascending Node

Geocentric Coordinates at Greatest Eclipse

	Sun	Moon
R.A. =	06h21m03.1s	18h21m02.6s
Dec. =	+23°20'50.2"	-23°20'06.9"
S.D. =	00°15'44.1"	00°16'00.4"
H.P. =	00°00'08.7"	00°58'44.7"

Eclipse Durations
Penumbral = 05h36m02s
Umbral = 03h40m21s
Total = 01h42m41s

Eclipse Contacts
P1 = 00:34:11 UT1
U1 = 01:31:57 UT1
U2 = 02:30:48 UT1
U3 = 04:13:29 UT1
U4 = 05:12:18 UT1
P4 = 06:10:13 UT1

Parameters
Eph. = JPL DE430
Rule = Herald-Sinnott
ΔT = 74 s

P4 U4 U3 U2 U1 P1
Eclipse at MoonRise
All Eclipse Visible
Eclipse at MoonSet
No Eclipse Visible

Moon in Zenith at Greatest Eclipse

No Eclipse Visible In Shaded Region

www.EclipseWise.com   ©2020 F. Espenak

Eclipse During Moon Rise

P1 - Penumbral Eclipse Begins
U1 - Partial Eclipse Begins
U2 - Total Eclipse Begins
U3 - Total Eclipse Ends
U4 - Partial Eclipse Ends
P4 - Penumbral Eclipse Ends

Eclipse During Moon Set

P1 - Penumbral Eclipse Begins
U1 - Partial Eclipse Begins
U2 - Total Eclipse Begins
U3 - Total Eclipse Ends
U4 - Partial Eclipse Ends
P4 - Penumbral Eclipse Ends

Layout of Lunar Eclipse Figure

Upper Left: Moon's Path Thru Earth's Shadows
Lower Left: Map of Eclipse Visibility
Right Side: Parameters for Lunar Eclipse

Explanation of Parameters Used in Lunar Eclipse Figures

Greatest Eclipse – The instant when the Moon passes closest to the axis of Earth's shadow cone (Universal Time[9])

Penumbral Eclipse Magnitude – Fraction of the Moon's diameter immersed in the penumbra at greatest eclipse.

Umbral Eclipse Magnitude – The fraction of the Moon's diameter immersed in the umbra at greatest eclipse.

Gamma – Minimum distance from the Moon's center to Earth's shadow axis (units of Earth's equatorial radius).

Axis – Minimum distance from the Moon's center to Earth's shadow axis (units of degrees).

Saros Series – The Saros series that the eclipse belongs to. The numbers in "[ ]" are the eclipse's sequential position and the number of eclipses in the Saros series.

Node – The orbital node near which the eclipse takes place (Ascending Node or Descending Node).

Geocentric Coordinates of the Sun and the Moon at Greatest Eclipse

      R.A. – Right Ascension   **S.D.** – Semi-Diameter (i.e. - radius)

      Dec. – Declination   **H.P.** – Horizontal Parallax

Eclipse Durations – Durations of Penumbral, Partial, and Total Eclipse.

Eclipse Contacts – Contact Times (Universal Time or UT1) of the Moon with the Penumbra and the Umbra

      P1, P4 – Start and End of the Penumbral Eclipse

      U1, U4 – Start and End of the Partial Eclipse

      U2, U3 – Start and End of the Total Eclipse

---

[9] Universal Time or UT1 is the modern replacement for Greenwich Mean Time

Penumbral Lunar Eclipse
2031 May 07

Greatest Eclipse = 03:50:46.7 UT1
Penumbral Eclipse Magnitude = 0.8827
Umbral Eclipse Magnitude = -0.0892
Gamma = -1.0695
Axis = 1.0670°
Saros = 112 [66 of 72]
Ascending Node

Geocentric Coordinates at Greatest Eclipse

Sun	Moon
R.A. = 02h55m49.7s	R.A. = 14h54m58.0s
Dec. = +16°44'40.2"	Dec. = -17°47'29.4"
S.D. = 00°15'51.2"	S.D. = 00°16'18.7"
H.P. = 00°00'08.7"	H.P. = 00°59'52.0"

Eclipse Durations	Eclipse Contacts
Penumbral = 03h58m04s	P1 = 01:51:42 UT1
	P4 = 05:49:46 UT1

Parameters
Eph. = JPL DE430
Rule = Herald-Sinnott
ΔT = 75 s

www.EclipseWise.com   ©2020 F. Espenak

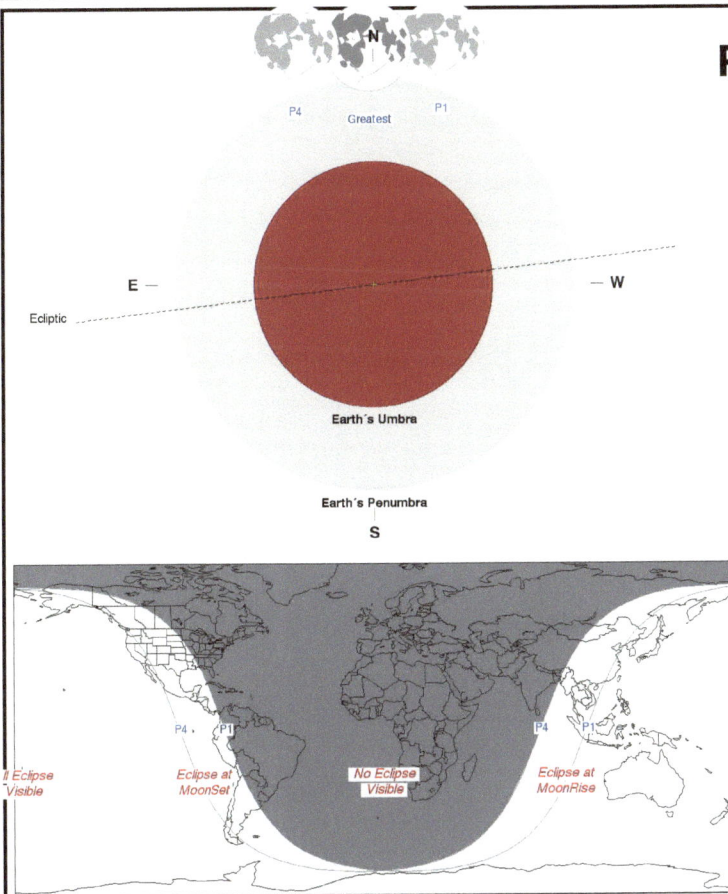

Penumbral Lunar Eclipse
2031 Jun 05

Greatest Eclipse = 11:44:02.8 UT1
Penumbral Eclipse Magnitude = 0.1306
Umbral Eclipse Magnitude = -0.8185
Gamma = 1.4732
Axis = 1.4967°
Saros = 150 [2 of 71]
Ascending Node

Geocentric Coordinates at Greatest Eclipse

Sun	Moon
R.A. = 04h53m21.6s	R.A. = 16h53m29.4s
Dec. = +22°33'01.5"	Dec. = -21°03'14.0"
S.D. = 00°15'45.9"	S.D. = 00°16'36.6"
H.P. = 00°00'08.7"	H.P. = 01°00'57.7"

Eclipse Durations	Eclipse Contacts
Penumbral = 01h36m18s	P1 = 10:55:58 UT1
	P4 = 12:32:17 UT1

Parameters
Eph. = JPL DE430
Rule = Herald-Sinnott
ΔT = 75 s

www.EclipseWise.com   ©2020 F. Espenak

Penumbral Lunar Eclipse
2031 Oct 30

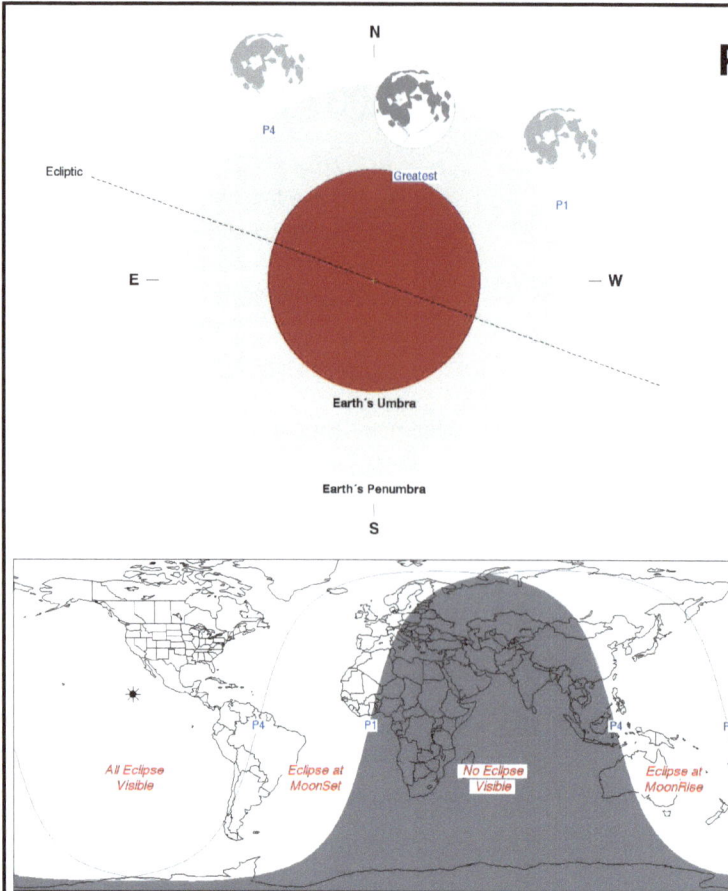

Greatest Eclipse = 07:45:29.4 UT1
Penumbral Eclipse Magnitude = 0.7173
Umbral Eclipse Magnitude = -0.3193
Gamma = 1.1774
Axis = 1.1188°
Saros = 117 [53 of 71]
Decending Node

Geocentric Coordinates at Greatest Eclipse

Sun	Moon
R.A. = 14h17m25.0s	R.A. = 02h16m19.7s
Dec. = -13°44'38.7"	Dec. = +14°49'53.3"
S.D. = 00°16'06.3"	S.D. = 00°15'32.2"
H.P. = 00°00'08.9"	H.P. = 00°57'01.3"

Eclipse Durations	Eclipse Contacts
Penumbral = 03h52m28s	P1 = 05:49:04 UT1
	P4 = 09:41:32 UT1

Parameters
Eph. = JPL DE430
Rule = Herald-Sinnott
ΔT = 75 s

www.EclipseWise.com ©2020 F. Espenak

Total Lunar Eclipse
2032 Apr 25

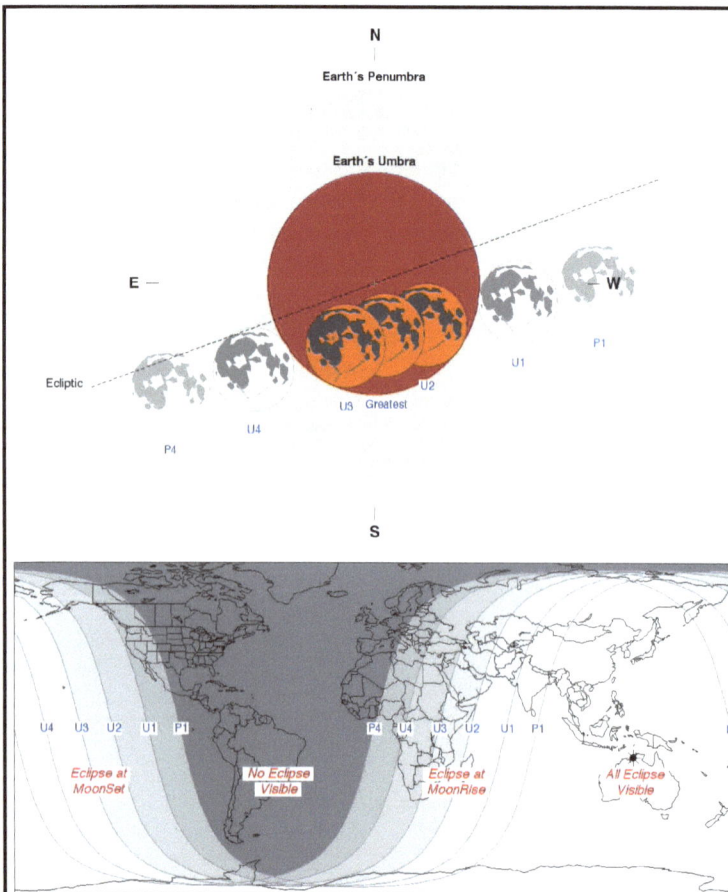

Greatest Eclipse = 15:13:35.5 UT1
Penumbral Eclipse Magnitude = 2.2204
Umbral Eclipse Magnitude = 1.1925
Gamma = -0.3558
Axis = 0.3365°
Saros = 122 [57 of 74]
Ascending Node

Geocentric Coordinates at Greatest Eclipse

Sun	Moon
R.A. = 02h14m38.2s	R.A. = 14h14m18.6s
Dec. = +13°30'28.8"	Dec. = -13°50'06.1"
S.D. = 00°15'53.8"	S.D. = 00°15'27.9"
H.P. = 00°00'08.7"	H.P. = 00°56'45.4"

Eclipse Durations	Eclipse Contacts
Penumbral = 05h43m18s	P1 = 12:21:52 UT1
Umbral = 03h31m58s	U1 = 13:27:35 UT1
Total = 01h06m10s	U2 = 14:40:24 UT1
	U3 = 15:46:34 UT1
	U4 = 16:59:33 UT1
	P4 = 18:05:10 UT1

Parameters
Eph. = JPL DE430
Rule = Herald-Sinnott
ΔT = 75 s

www.EclipseWise.com ©2020 F. Espenak

Total Lunar Eclipse
2032 Oct 18

Greatest Eclipse = 19:02:24.9 UT1

Penumbral Eclipse Magnitude = 2.0841

Umbral Eclipse Magnitude = 1.1039

Gamma = 0.4169

Axis = 0.4177°

Saros = 127 [43 of 72]

Decending Node

Geocentric Coordinates at Greatest Eclipse

Sun	Moon
R.A. = 13h36m15.4s	R.A. = 01h35m47.9s
Dec. = -10°01'20.8"	Dec. = +10°25'28.7"
S.D. = 00°16'03.4"	S.D. = 00°16'22.8"
H.P. = 00°00'08.8"	H.P. = 01°00'07.0"

Eclipse Durations
Penumbral = 05h16m13s
Umbral = 03h16m38s
Total = 00h47m40s

Eclipse Contacts
P1 = 16:24:17 UT1
U1 = 17:24:00 UT1
U2 = 18:38:24 UT1
U3 = 19:26:04 UT1
U4 = 20:40:38 UT1
P4 = 21:40:30 UT1

Parameters
Eph. = JPL DE430
Rule = Herald-Sinnott
ΔT = 75 s

www.EclipseWise.com   ©2020 F. Espenak

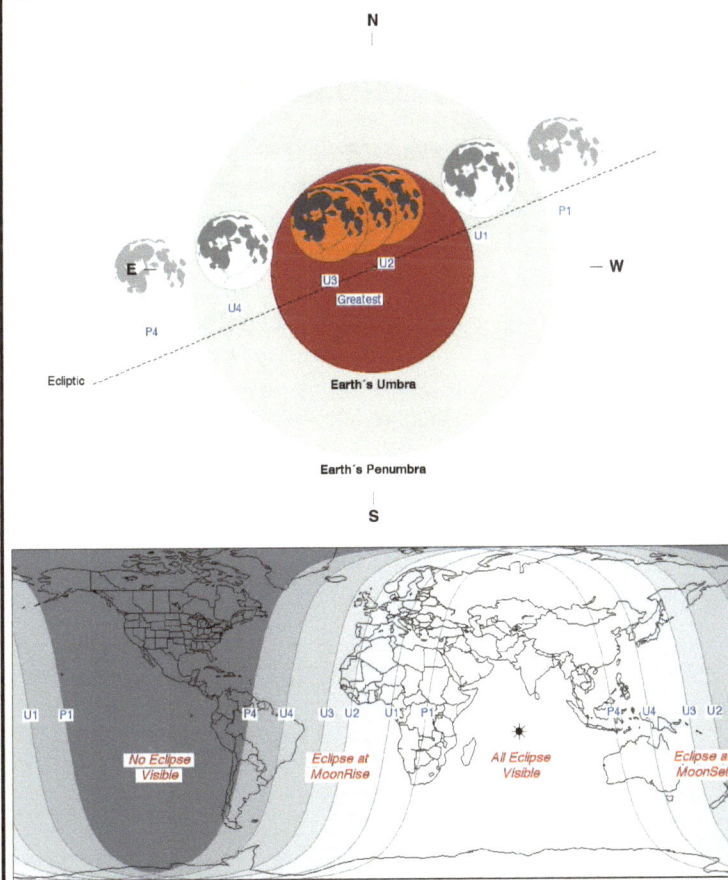

Total Lunar Eclipse
2033 Apr 14

Greatest Eclipse = 19:12:35.9 UT1

Penumbral Eclipse Magnitude = 2.1722

Umbral Eclipse Magnitude = 1.0955

Gamma = 0.3954

Axis = 0.3581°

Saros = 132 [31 of 71]

Ascending Node

Geocentric Coordinates at Greatest Eclipse

Sun	Moon
R.A. = 01h33m13.7s	R.A. = 13h33m37.2s
Dec. = +09°43'50.2"	Dec. = -09°23'08.7"
S.D. = 00°15'56.7"	S.D. = 00°14'48.5"
H.P. = 00°00'08.8"	H.P. = 00°54'21.0"

Eclipse Durations
Penumbral = 06h02m05s
Umbral = 03h35m47s
Total = 00h49m50s

Eclipse Contacts
P1 = 16:11:35 UT1
U1 = 17:24:48 UT1
U2 = 18:47:53 UT1
U3 = 19:37:42 UT1
U4 = 21:00:35 UT1
P4 = 22:13:40 UT1

Parameters
Eph. = JPL DE430
Rule = Herald-Sinnott
ΔT = 76 s

www.EclipseWise.com   ©2020 F. Espenak

Total Lunar Eclipse
2033 Oct 08

Greatest Eclipse = 10:55:06.9 UT1
Penumbral Eclipse Magnitude = 2.3068
Umbral Eclipse Magnitude = 1.3508
Gamma = -0.2889
Axis = 0.2958°
Saros = 137 [27 of 78]
Decending Node

Geocentric Coordinates at Greatest Eclipse

Sun	Moon
R.A. = 12h57m01.9s	R.A. = 00h57m22.8s
Dec. = -06°05'34.4"	Dec. = +05°48'36.0"
S.D. = 00°16'00.5"	S.D. = 00°16'44.6"
H.P. = 00°00'08.8"	H.P. = 01°01'27.1"

Eclipse Durations	Eclipse Contacts
Penumbral = 05h13m25s	P1 = 08:18:27 UT1
Umbral = 03h23m07s	U1 = 09:13:37 UT1
Total = 01h19m25s	U2 = 10:15:31 UT1
	U3 = 11:34:56 UT1
	U4 = 12:36:43 UT1
	P4 = 13:31:51 UT1

Parameters
Eph. = JPL DE430
Rule = Herald-Sinnott
ΔT =   76 s

www.EclipseWise.com   ©2020 F. Espenak

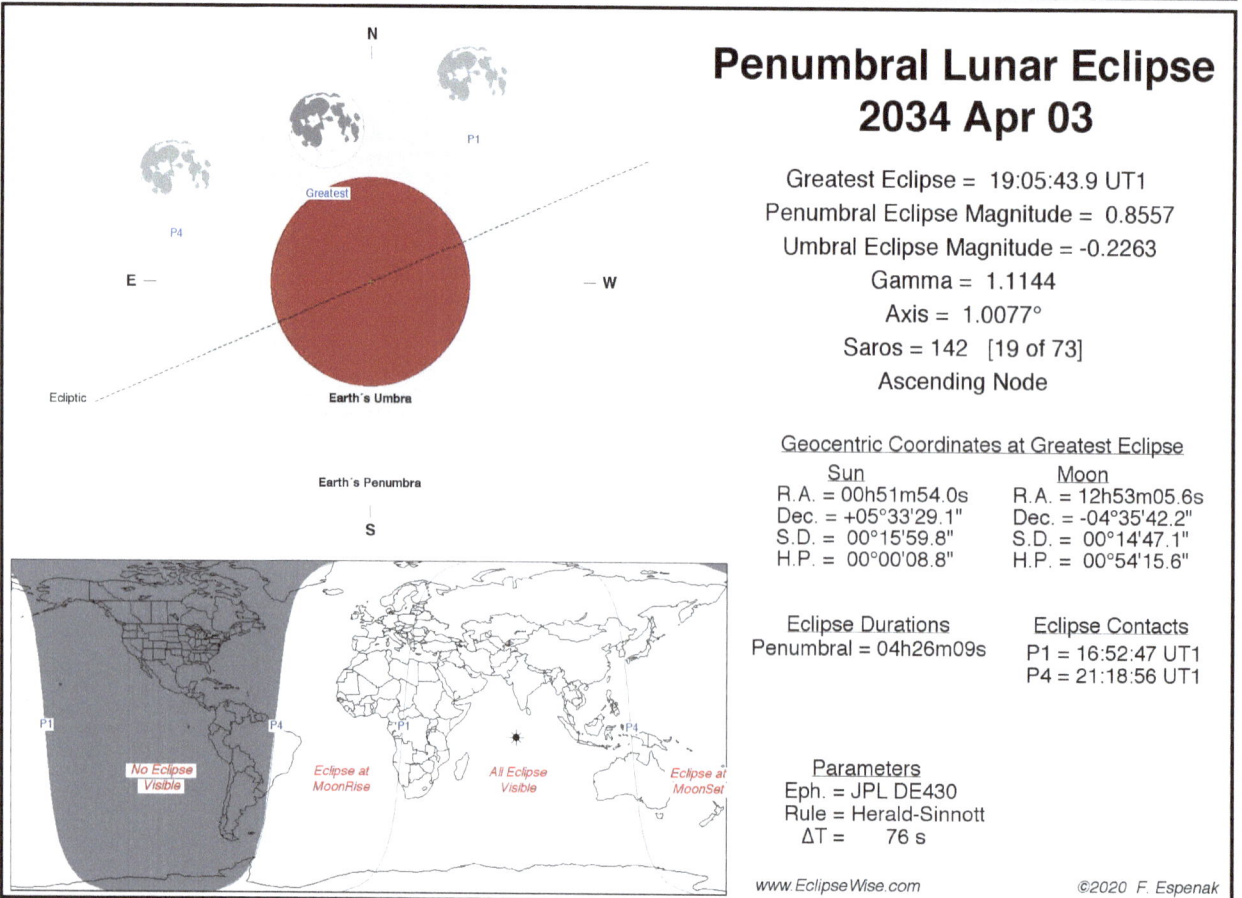

Penumbral Lunar Eclipse
2034 Apr 03

Greatest Eclipse = 19:05:43.9 UT1
Penumbral Eclipse Magnitude = 0.8557
Umbral Eclipse Magnitude = -0.2263
Gamma = 1.1144
Axis = 1.0077°
Saros = 142 [19 of 73]
Ascending Node

Geocentric Coordinates at Greatest Eclipse

Sun	Moon
R.A. = 00h51m54.0s	R.A. = 12h53m05.6s
Dec. = +05°33'29.1"	Dec. = -04°35'42.2"
S.D. = 00°15'59.8"	S.D. = 00°14'47.1"
H.P. = 00°00'08.8"	H.P. = 00°54'15.6"

Eclipse Durations	Eclipse Contacts
Penumbral = 04h26m09s	P1 = 16:52:47 UT1
	P4 = 21:18:56 UT1

Parameters
Eph. = JPL DE430
Rule = Herald-Sinnott
ΔT =   76 s

www.EclipseWise.com   ©2020 F. Espenak

N
Earth's Penumbra

Ecliptic

Earth's Umbra

E —   — W

P4

U4 U1
Greatest   S

Partial Lunar Eclipse
2034 Sep 28

Greatest Eclipse = 02:46:21.1 UT1

Penumbral Eclipse Magnitude = 0.9922

Umbral Eclipse Magnitude = 0.0155

Gamma = -1.0110

Axis = 1.0104°

Saros = 147 [9 of 70]

Decending Node

Geocentric Coordinates at Greatest Eclipse

Sun	Moon
R.A. = 12h18m35.8s	R.A. = 00h19m50.0s
Dec. = -02°00'43.0"	Dec. = +01°02'59.4"
S.D. = 00°15'57.6"	S.D. = 00°16'20.4"
H.P. = 00°00'08.8"	H.P. = 00°59'58.2"

Eclipse Durations	Eclipse Contacts
Penumbral = 04h09m22s	P1 = 00:41:50 UT1
Umbral = 00h27m48s	U1 = 02:32:44 UT1
	U4 = 03:00:33 UT1
	P4 = 04:51:12 UT1

Parameters
Eph. = JPL DE430
Rule = Herald-Sinnott
ΔT =   76 s

P1

P4   U4 U1   P1   P4   U4 J1   P1

Eclipse at   All Eclipse   Eclipse at   No Eclipse
MoonRise   Visible   MoonSet   Visible

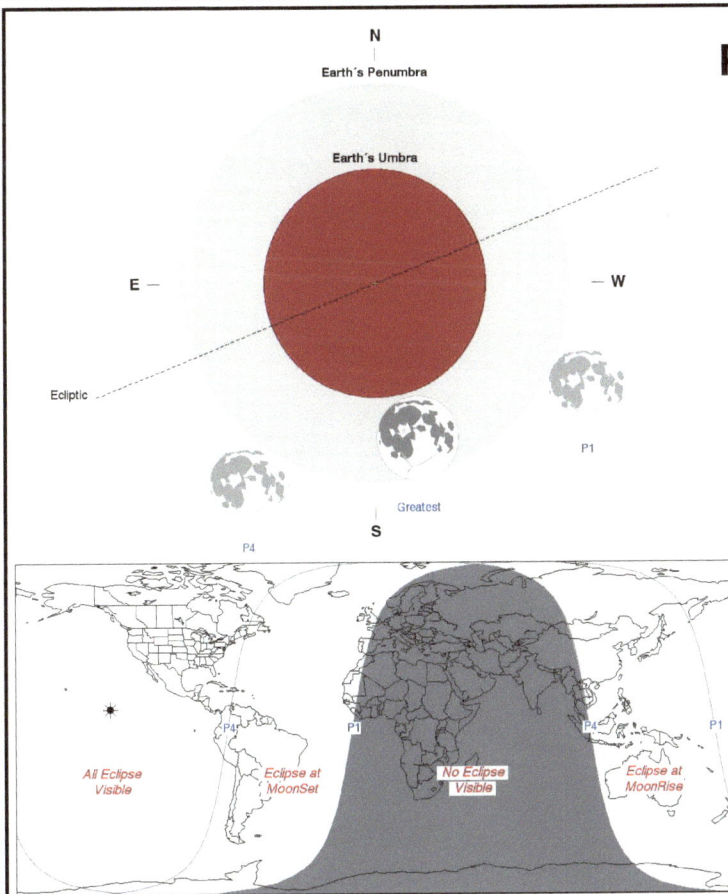

N
Earth's Penumbra

Earth's Umbra

E —   — W

Ecliptic

Greatest
S

Penumbral Lunar Eclipse
2035 Feb 22

Greatest Eclipse = 09:04:55.0 UT1

Penumbral Eclipse Magnitude = 0.9663

Umbral Eclipse Magnitude = -0.0523

Gamma = -1.0367

Axis = 1.0066°

Saros = 114 [60 of 71]

Ascending Node

Geocentric Coordinates at Greatest Eclipse

Sun	Moon
R.A. = 22h21m54.2s	R.A. = 10h20m48.3s
Dec. = -10°11'53.9"	Dec. = +09°13'43.5"
S.D. = 00°16'10.2"	S.D. = 00°15'52.5"
H.P. = 00°00'08.9"	H.P. = 00°58'15.8"

Eclipse Durations	Eclipse Contacts
Penumbral = 04h16m23s	P1 = 06:56:33 UT1
	P4 = 11:12:57 UT1

P1

P4

P4   P1   No Eclipse   Eclipse at
                         Visible   MoonRise

All Eclipse   Eclipse at
Visible   MoonRise

Parameters
Eph. = JPL DE430
Rule = Herald-Sinnott
ΔT =   76 s

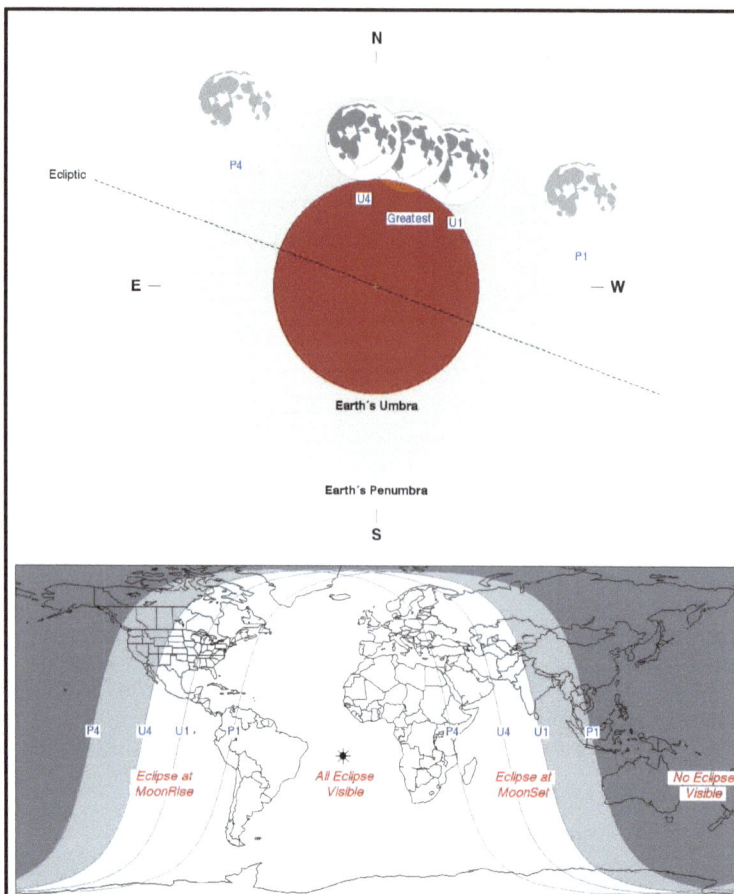

Partial Lunar Eclipse
2035 Aug 19

Greatest Eclipse = 01:10:58.2 UT1
Penumbral Eclipse Magnitude = 1.1519
Umbral Eclipse Magnitude = 0.1049
Gamma = 0.9434
Axis = 0.8708°
Saros = 119 [62 of 82]
Decending Node

Geocentric Coordinates at Greatest Eclipse

Sun	Moon
R.A. = 09h52m42.8s	R.A. = 21h51m50.7s
Dec. = +12°52'21.3"	Dec. = -12°01'40.6"
S.D. = 00°15'48.1"	S.D. = 00°15'05.5"
H.P. = 00°00'08.7"	H.P. = 00°55'23.4"

Eclipse Durations	Eclipse Contacts
Penumbral = 04h50m38s	P1 = 22:45:34 UT1
Umbral = 01h17m11s	U1 = 00:32:18 UT1
	U4 = 01:49:29 UT1
	P4 = 03:36:13 UT1

Parameters
Eph. = JPL DE430
Rule = Herald-Sinnott
ΔT = 77 s

www.EclipseWise.com   ©2020 F. Espenak

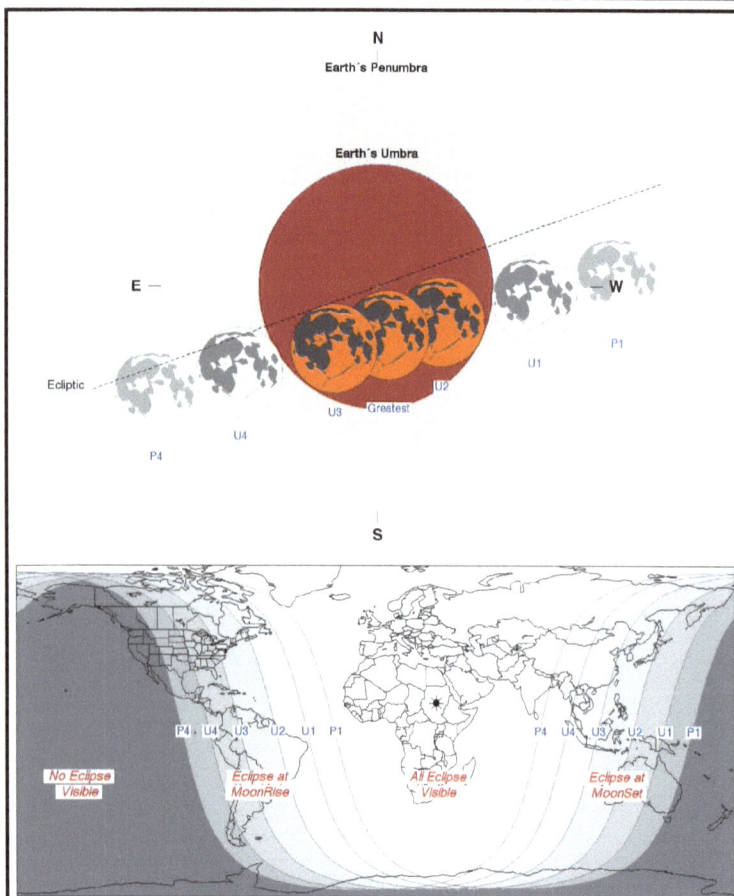

Total Lunar Eclipse
2036 Feb 11

Greatest Eclipse = 22:11:49.2 UT1
Penumbral Eclipse Magnitude = 2.2762
Umbral Eclipse Magnitude = 1.3006
Gamma = -0.3110
Axis = 0.3159°
Saros = 124 [50 of 73]
Ascending Node

Geocentric Coordinates at Greatest Eclipse

Sun	Moon
R.A. = 21h40m25.4s	R.A. = 09h40m07.3s
Dec. = -13°55'30.0"	Dec. = +13°37'03.4"
S.D. = 00°16'12.3"	S.D. = 00°16'36.7"
H.P. = 00°00'08.9"	H.P. = 01°00'57.8"

Eclipse Durations	Eclipse Contacts
Penumbral = 05h16m53s	P1 = 19:33:22 UT1
Umbral = 03h22m39s	U1 = 20:30:26 UT1
Total = 01h15m06s	U2 = 21:34:10 UT1
	U3 = 22:49:16 UT1
	U4 = 23:53:05 UT1
	P4 = 00:50:15 UT1

Parameters
Eph. = JPL DE430
Rule = Herald-Sinnott
ΔT = 77 s

www.EclipseWise.com   ©2020 F. Espenak

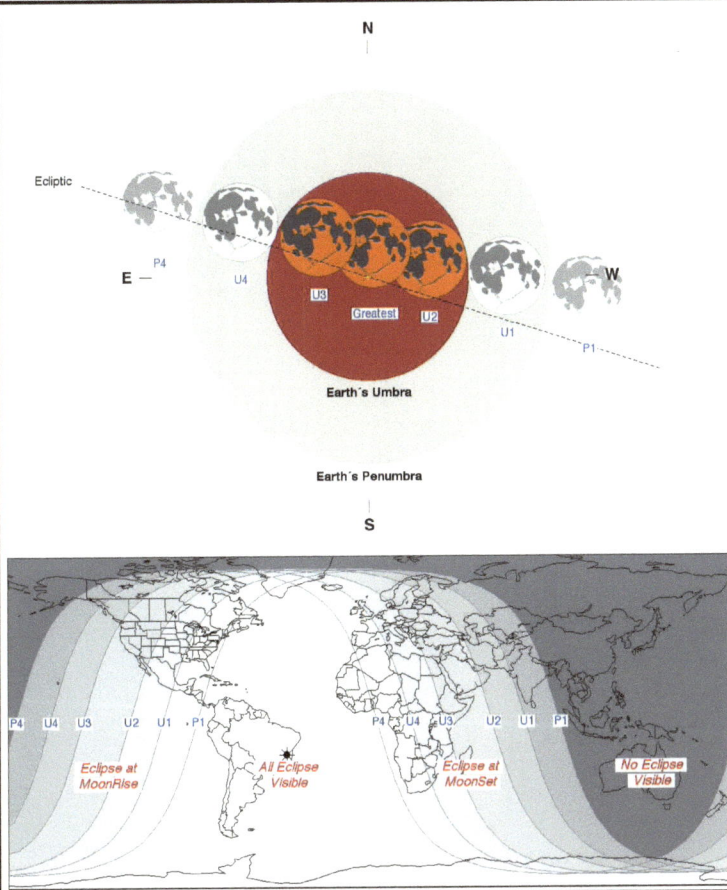

Total Lunar Eclipse
2036 Aug 07

Greatest Eclipse = 02:51:15.2 UT1
Penumbral Eclipse Magnitude = 2.5279
Umbral Eclipse Magnitude = 1.4556
Gamma = 0.2004
Axis = 0.1803°
Saros = 129 [39 of 71]
Decending Node

Geocentric Coordinates at Greatest Eclipse

Sun	Moon
R.A. = 09h10m39.1s	R.A. = 21h10m30.3s
Dec. = +16°16'20.8"	Dec. = -16°05'44.3"
S.D. = 00°15'46.3"	S.D. = 00°14'42.5"
H.P. = 00°00'08.7"	H.P. = 00°53'58.8"

Eclipse Durations
Penumbral = 06h13m03s
Umbral = 03h52m11s
Total = 01h36m06s

Eclipse Contacts
P1 = 23:44:42 UT1
U1 = 00:55:08 UT1
U2 = 02:03:08 UT1
U3 = 03:39:15 UT1
U4 = 04:47:19 UT1
P4 = 05:57:45 UT1

Parameters
Eph. = JPL DE430
Rule = Herald-Sinnott
ΔT = 77 s

www.EclipseWise.com   ©2020 F. Espenak

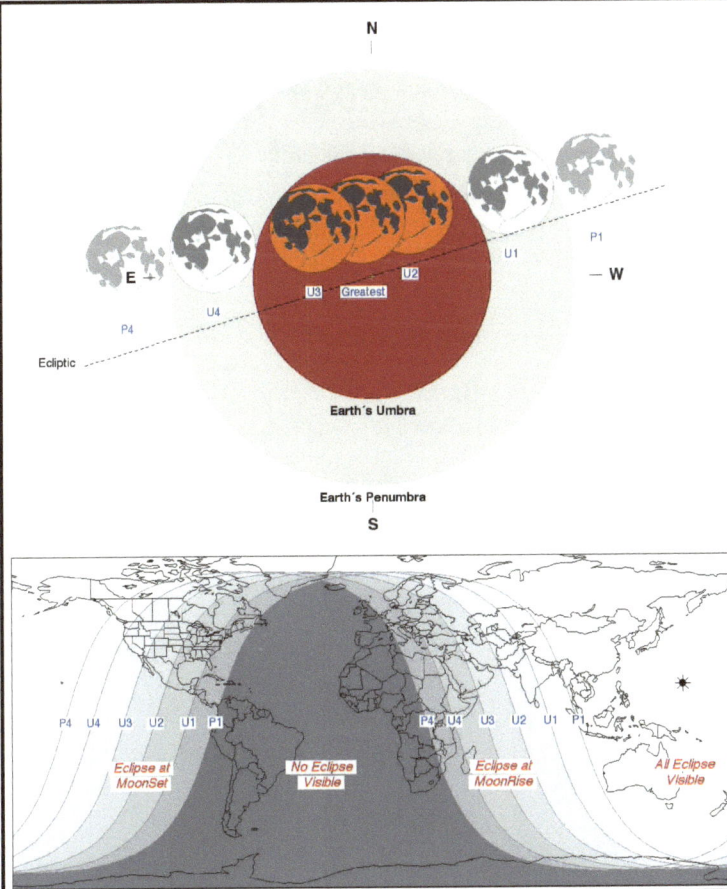

Total Lunar Eclipse
2037 Jan 31

Greatest Eclipse = 14:00:20.9 UT1
Penumbral Eclipse Magnitude = 2.1815
Umbral Eclipse Magnitude = 1.2086
Gamma = 0.3619
Axis = 0.3693°
Saros = 134 [28 of 72]
Ascending Node

Geocentric Coordinates at Greatest Eclipse

Sun	Moon
R.A. = 20h57m58.6s	R.A. = 08h58m15.6s
Dec. = -17°10'47.4"	Dec. = +17°32'34.5"
S.D. = 00°16'14.0"	S.D. = 00°16'41.1"
H.P. = 00°00'08.9"	H.P. = 01°01'14.2"

Eclipse Durations
Penumbral = 05h12m56s
Umbral = 03h18m12s
Total = 01h04m17s

Eclipse Contacts
P1 = 11:23:54 UT1
U1 = 12:21:18 UT1
U2 = 13:28:18 UT1
U3 = 14:32:35 UT1
U4 = 15:39:30 UT1
P4 = 16:36:50 UT1

Parameters
Eph. = JPL DE430
Rule = Herald-Sinnott
ΔT = 77 s

www.EclipseWise.com   ©2020 F. Espenak

N

Earth's Penumbra

Earth's Umbra

Ecliptic

E —   — W

P4   U4   U1   P1

Greatest

S

Partial Lunar Eclipse
2037 Jul 27

Greatest Eclipse = 04:08:35.6 UT1
Penumbral Eclipse Magnitude = 1.8596
Umbral Eclipse Magnitude = 0.8108
Gamma = -0.5582
Axis = 0.5126°
Saros = 139 [22 of 79]
Decending Node

Geocentric Coordinates at Greatest Eclipse

Sun	Moon
R.A. = 08h27m18.9s	R.A. = 20h27m37.3s
Dec. = +19°07'58.8"	Dec. = -19°38'25.9"
S.D. = 00°15'45.0"	S.D. = 00°15'00.9"
H.P. = 00°00'08.7"	H.P. = 00°55'06.5"

Eclipse Durations	Eclipse Contacts
Penumbral = 05h41m44s	P1 = 01:17:47 UT1
Umbral = 03h13m11s	U1 = 02:32:01 UT1
	U4 = 05:45:12 UT1
	P4 = 06:59:31 UT1

Parameters
Eph. = JPL DE430
Rule = Herald-Sinnott
ΔT = 78 s

U4   U1   P1   P4 U4   U1 P1   P4

Eclipse at   All Eclipse   Eclipse at   No Eclipse
MoonRise   Visible   MoonSet   Visible

www.EclipseWise.com   ©2020 F. Espenak

N

P1

P4

Greatest

E —   — W

Ecliptic

Earth's Umbra

Earth's Penumbra

S

Penumbral Lunar Eclipse
2038 Jan 21

Greatest Eclipse = 03:48:33.9 UT1
Penumbral Eclipse Magnitude = 0.9009
Umbral Eclipse Magnitude = -0.1127
Gamma = 1.0711
Axis = 1.0505°
Saros = 144 [17 of 71]
Ascending Node

Geocentric Coordinates at Greatest Eclipse

Sun	Moon
R.A. = 20h13m39.3s	R.A. = 08h14m12.5s
Dec. = -19°53'23.0"	Dec. = +20°55'55.8"
S.D. = 00°16'15.2"	S.D. = 00°16'02.1"
H.P. = 00°00'08.9"	H.P. = 00°58'51.1"

Eclipse Durations	Eclipse Contacts
Penumbral = 04h06m32s	P1 = 01:45:24 UT1
	P4 = 05:51:56 UT1

Parameters
Eph. = JPL DE430
Rule = Herald-Sinnott
ΔT = 78 s

P4   P1   P4   P1

Eclipse at   All Eclipse   Eclipse at   No Eclipse
MoonRise   Visible   MoonSet   Visible

www.EclipseWise.com   ©2020 F. Espenak

Penumbral Lunar Eclipse
2038 Jun 17

Greatest Eclipse = 02:43:44.0 UT1
Penumbral Eclipse Magnitude = 0.4438
Umbral Eclipse Magnitude = -0.5259
Gamma = 1.3083
Axis = 1.2993°
Saros = 111 [68 of 71]
Decending Node

Geocentric Coordinates at Greatest Eclipse

Sun	Moon
R.A. = 05h42m46.1s	R.A. = 17h43m28.2s
Dec. = +23°22'28.6"	Dec. = -22°05'07.2"
S.D. = 00°15'44.7"	S.D. = 00°16'14.3"
H.P. = 00°00'08.7"	H.P. = 00°59'35.6"

Eclipse Durations	Eclipse Contacts
Penumbral = 02h57m03s	P1 = 01:15:11 UT1
	P4 = 04:12:14 UT1

Parameters
Eph. = JPL DE430
Rule = Herald-Sinnott
ΔT = 78 s

www.EclipseWise.com   ©2020 F. Espenak

Penumbral Lunar Eclipse
2038 Jul 16

Greatest Eclipse = 11:34:38.0 UT1
Penumbral Eclipse Magnitude = 0.5012
Umbral Eclipse Magnitude = -0.4938
Gamma = -1.2838
Axis = 1.2417°
Saros = 149 [4 of 71]
Decending Node

Geocentric Coordinates at Greatest Eclipse

Sun	Moon
R.A. = 07h43m47.7s	R.A. = 19h44m13.1s
Dec. = +21°17'34.6"	Dec. = -22°31'51.1"
S.D. = 00°15'44.2"	S.D. = 00°15'48.9"
H.P. = 00°00'08.7"	H.P. = 00°58'02.4"

Eclipse Durations	Eclipse Contacts
Penumbral = 03h13m09s	P1 = 09:58:00 UT1
	P4 = 13:11:09 UT1

Parameters
Eph. = JPL DE430
Rule = Herald-Sinnott
ΔT = 78 s

www.EclipseWise.com   ©2020 F. Espenak

43

Penumbral Lunar Eclipse
2038 Dec 11

N

Earth's Penumbra

Earth's Umbra

Ecliptic

E —

— W

P4

Greatest

S

P1

Greatest Eclipse = 17:43:41.9 UT1
Penumbral Eclipse Magnitude = 0.8062
Umbral Eclipse Magnitude = -0.2876
Gamma = -1.1449
Axis = 1.0398°
Saros = 116 [59 of 73]
Ascending Node

Geocentric Coordinates at Greatest Eclipse

Sun	Moon
R.A. = 17h15m29.9s	R.A. = 05h16m16.9s
Dec. = -23°02'24.2"	Dec. = +22°00'57.8"
S.D. = 00°16'14.6"	S.D. = 00°14'51.0"
H.P. = 00°00'08.9"	H.P. = 00°54'29.8"

Eclipse Durations	Eclipse Contacts
Penumbral = 04h19m16s	P1 = 15:34:11 UT1
	P4 = 19:53:27 UT1

P1

P4

P1

P4

clipse at
MoonSet

No Eclipse
Visible

Eclipse at
MoonRise

All Eclipse
Visible

Parameters
Eph. = JPL DE430
Rule = Herald-Sinnott
ΔT = 79 s

Partial Lunar Eclipse
2039 Jun 06

N

P1

P1

Ecliptic

E —

U4

Greatest

— W

P4

U1

Earth's Umbra

Earth's Penumbra

S

Greatest Eclipse = 18:53:06.7 UT1
Penumbral Eclipse Magnitude = 1.8288
Umbral Eclipse Magnitude = 0.8863
Gamma = 0.5460
Axis = 0.5584°
Saros = 121 [56 of 82]
Decending Node

Geocentric Coordinates at Greatest Eclipse

Sun	Moon
R.A. = 04h58m56.4s	R.A. = 16h59m25.6s
Dec. = +22°41'33.8"	Dec. = -22°08'44.6"
S.D. = 00°15'45.7"	S.D. = 00°16'43.4"
H.P. = 00°00'08.7"	H.P. = 01°01'22.4"

Eclipse Durations	Eclipse Contacts
Penumbral = 04h57m32s	P1 = 16:24:23 UT1
Umbral = 03h00m00s	U1 = 17:23:10 UT1
	U4 = 20:23:10 UT1
	P4 = 21:21:55 UT1

U1 P1

P4 U4

U1 P1

P4 U4

No Eclipse
Visible

Eclipse at
MoonRise

All Eclipse
Visible

Eclipse at
MoonSet

Parameters
Eph. = JPL DE430
Rule = Herald-Sinnott
ΔT = 79 s

Partial Lunar Eclipse
2039 Nov 30

Greatest Eclipse = 16:55:08.5 UT1
Penumbral Eclipse Magnitude = 2.0435
Umbral Eclipse Magnitude = 0.9443
Gamma = -0.4721
Axis = 0.4260°
Saros = 126 [46 of 70]
Ascending Node

Geocentric Coordinates at Greatest Eclipse

Sun	Moon
R.A. = 16h26m20.8s	R.A. = 04h26m48.9s
Dec. = -21°41'27.9"	Dec. = +21°16'45.4"
S.D. = 00°16'13.0"	S.D. = 00°14'45.3"
H.P. = 00°00'08.9"	H.P. = 00°54'08.9"

Eclipse Durations
Penumbral = 06h00m59s
Umbral = 03h26m48s

Eclipse Contacts
P1 = 13:54:43 UT1
U1 = 15:11:49 UT1
U4 = 18:38:37 UT1
P4 = 19:55:42 UT1

Parameters
Eph. = JPL DE430
Rule = Herald-Sinnott
ΔT = 79 s

www.EclipseWise.com ©2020 F. Espenak

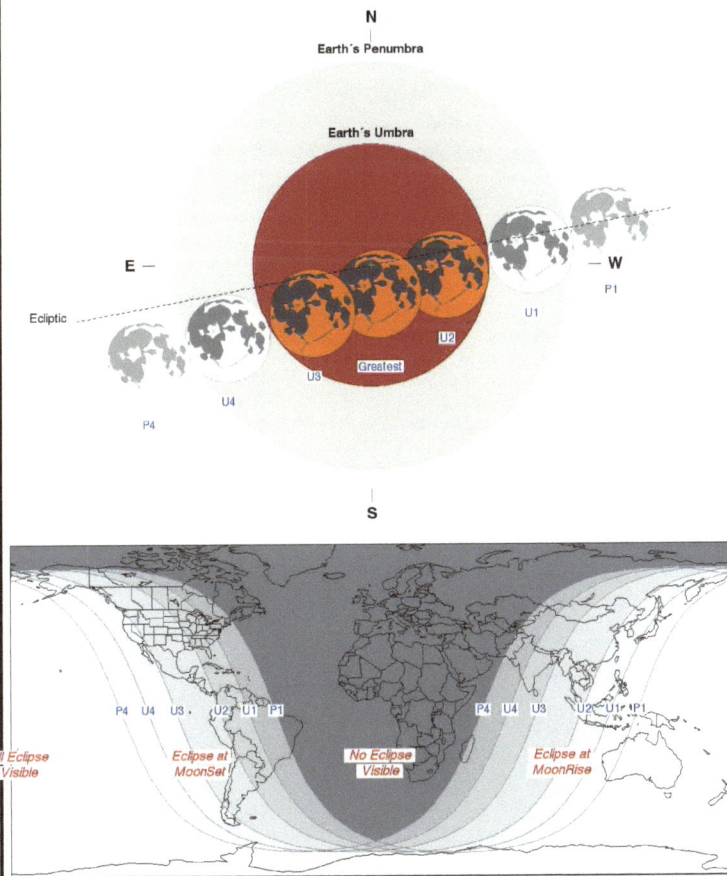

Total Lunar Eclipse
2040 May 26

Greatest Eclipse = 11:45:02.2 UT1
Penumbral Eclipse Magnitude = 2.4955
Umbral Eclipse Magnitude = 1.5365
Gamma = -0.1872
Axis = 0.1885°
Saros = 131 [35 of 72]
Decending Node

Geocentric Coordinates at Greatest Eclipse

Sun	Moon
R.A. = 04h15m46.6s	R.A. = 16h15m33.4s
Dec. = +21°16'35.1"	Dec. = -21°27'28.2"
S.D. = 00°15'47.2"	S.D. = 00°16'27.7"
H.P. = 00°00'08.7"	H.P. = 01°00'24.9"

Eclipse Durations
Penumbral = 05h22m11s
Umbral = 03h31m28s
Total = 01h32m56s

Eclipse Contacts
P1 = 09:03:54 UT1
U1 = 09:59:17 UT1
U2 = 10:58:31 UT1
U3 = 12:31:27 UT1
U4 = 13:30:45 UT1
P4 = 14:26:05 UT1

Parameters
Eph. = JPL DE430
Rule = Herald-Sinnott
ΔT = 79 s

www.EclipseWise.com ©2020 F. Espenak

N

Ecliptic

P4

E —

U4

U3

U1

Greatest

U2

U1

— W

P1

Earth's Umbra

Earth's Penumbra

S

Total Lunar Eclipse
2040 Nov 18

Greatest Eclipse = 19:03:20.9 UT1

Penumbral Eclipse Magnitude = 2.4543

Umbral Eclipse Magnitude = 1.3991

Gamma = 0.2361

Axis = 0.2215°

Saros = 136   [21 of 72]

Ascending Node

Geocentric Coordinates at Greatest Eclipse

Sun	Moon
R.A. = 15h39m03.9s	R.A. = 03h38m45.6s
Dec. = -19°29'49.7"	Dec. = +19°42'23.6"
S.D. = 00°16'11.0"	S.D. = 00°15'20.2"
H.P. = 00°00'08.9"	H.P. = 00°56'17.3"

Eclipse Durations

Penumbral = 05h54m26s

Umbral = 03h41m13s

Total = 01h28m33s

Eclipse Contacts

P1 = 16:06:09 UT1
U1 = 17:12:40 UT1
U2 = 18:18:57 UT1
U3 = 19:47:30 UT1
U4 = 20:53:53 UT1
P4 = 22:00:35 UT1

Parameters

Eph. = JPL DE430
Rule = Herald-Sinnott
ΔT =   80 s

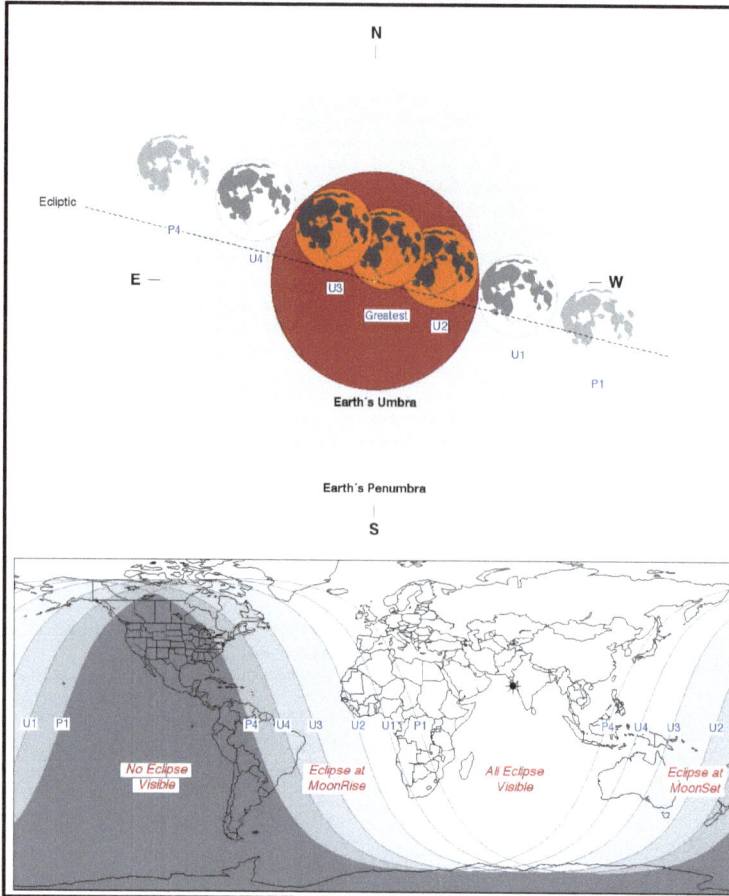

U1 P1

No Eclipse
Visible

P4 U4 U3 U2 U1 P1

Eclipse at
MoonRise

All Eclipse
Visible

P4 U4 U3 U2

Eclipse at
MoonSet

www.EclipseWise.com   *©2020 F. Espenak*

Key to Catalog of Lunar Eclipses

The following catalog lists all lunar eclipses during a 25-year period.

A brief description of each parameter in the catalog appears below.

Date — Gregorian date of Greatest Eclipse

Greatest Eclipse — Universal Time of Greatest Eclipse

Saros— Saros Series Number of the eclipse

Type — Lunar Eclipse Type

> N = Penumbral Lunar Eclipse
> P = Partial Lunar Eclipse
> T = Total Lunar Eclipse
> m = Middle eclipse of Saros series
> + = Central total eclipse (Moon's center passes north of shadow axis)
> – = Central total eclipse (Moon's center passes south of shadow axis)
> * = Total penumbral eclipse
> b = Saros series begins (first penumbral eclipse in a Saros series)
> e = Saros series ends (last penumbral eclipse in a Saros series)

Gamma — minimum distance from center of the Moon to axis of Earth's umbral shadow

P. Mag — Penumbral Magnitude; fraction of the Moon's diameter immersed in the penumbra

U. Mag — Umbral Magnitude; fraction of the Moon's diameter immersed in the umbra

Phase Durations
> **Pen** — elapsed time from contact P1 to P4 (minutes)
> **Par** — elapsed time from contact U1 to U4 (minutes)
> **Total** — elapsed time from contact U2 to U3 (minutes)

Lat & Long — latitude and longitude where the Moon appears in the zenith at Greatest Eclipse

Photo 2–8 Time sequence of the total lunar eclipse of 2014 April 15. ©2014 F. Espenak

Catalog of Lunar Eclipses: 2031 to 2055

Date	Greatest Eclipse	Saros	Type	Gamma	P. Mag	U.Mag	Pen	Partial	Total	Lat.	Long.
2031 May 07	3:50:47	112	N	-1.0695	0.8827	-0.0892	238.1	-	-	17.8S	58.8W
2031 Jun 05	11:44:03	150	N	1.4732	0.1306	-0.8185	96.3	-	-	21.1S	176.3W
2031 Oct 30	7:45:29	117	N	1.1774	0.7173	-0.3193	232.5		-	14.8N	120.7W
2032 Apr 25	15:13:36	122	T	-0.3558	2.2204	1.1925	343.3	212.0	66.2	13.8S	131.0E
2032 Oct 18	19:02:25	127	T	0.4169	2.0841	1.1039	316.2	196.6	47.7	10.4N	70.5E
2033 Apr 14	19:12:36	132	T	0.3954	2.1722	1.0955	362.1	215.8	49.8	9.4S	72.0E
2033 Oct 08	10:55:07	137	T	-0.2889	2.3068	1.3508	313.4	203.1	79.4	5.8N	166.8W
2034 Apr 03	19:05:44	142	N	1.1144	0.8557	-0.2263	266.2	-	-	4.6S	74.7E
2034 Sep 28	2:46:21	147	P	-1.0110	0.9922	0.0155	249.4	27.8	-	1.0N	43.6W
2035 Feb 22	9:04:55	114	N	-1.0367	0.9663	-0.0523	256.4	-	-	9.2N	133.1W
2035 Aug 19	1:10:58	119	P	0.9434	1.1519	0.1049	290.6	77.2	-	12.0S	17.0W
2036 Feb 11	22:11:49	124	T	-0.3110	2.2762	1.3006	316.9	202.7	75.1	13.6N	30.5E
2036 Aug 07	2:51:15	129	T+	0.2004	2.5279	1.4556	373.1	232.2	96.1	16.1S	41.4W
2037 Jan 31	14:00:21	134	T	0.3619	2.1815	1.2086	312.9	198.2	64.3	17.5N	153.3E
2037 Jul 27	4:08:36	139	P	-0.5582	1.8596	0.8108	341.7	193.2	-	19.6S	60.4W
2038 Jan 21	3:48:34	144	N	1.0711	0.9009	-0.1127	246.5	-	-	20.9N	54.2W
2038 Jun 17	2:43:44	111	N	1.3083	0.4438	-0.5259	177.0	-	-	22.1S	40.5W
2038 Jul 16	11:34:38	149	N	-1.2838	0.5012	-0.4938	193.2	-	-	22.5S	172.0W
2038 Dec 11	17:43:42	116	N	-1.1449	0.8062	-0.2876	259.3	-	-	22.0N	92.6E
2039 Jun 06	18:53:07	121	P	0.5460	1.8288	0.8863	297.5	180.0	-	22.1S	76.5E
2039 Nov 30	16:55:09	126	P	-0.4721	2.0435	0.9443	361.0	206.8	-	21.3N	103.5E
2040 May 26	11:45:02	131	T-	-0.1872	2.4955	1.5365	322.2	211.5	92.9	21.5S	177.0W
2040 Nov 18	19:03:21	136	T+	0.2361	2.4543	1.3991	354.4	221.2	88.6	19.7N	70.4E
2041 May 16	0:41:42	141	P	-0.9747	1.0765	0.0663	270.5	59.4	-	20.0S	11.7W
2041 Nov 08	4:33:44	146	P	0.9212	1.1675	0.1714	268.7	91.1	-	17.5N	72.9W
2042 Apr 05	14:28:52	113	N	1.1080	0.8700	-0.2156	269.2	-	-	5.4S	143.9E
2042 Sep 29	10:44:27	118	N	-1.0262	0.9548	-0.0010	239.2	-	-	1.6N	163.0W
2043 Mar 25	14:30:43	123	T	0.3849	2.1920	1.1161	360.0	215.3	54.1	1.6S	144.0E
2043 Sep 19	1:50:29	128	T	-0.3316	2.2452	1.2575	326.4	206.7	72.3	1.9S	29.0W
2044 Mar 13	19:37:11	133	T	-0.3496	2.2322	1.2050	339.1	209.7	67.0	2.1N	67.8E
2044 Sep 07	11:19:23	138	T	0.4318	2.0879	1.0476	344.8	206.9	34.8	5.4S	170.6W
2045 Mar 03	7:42:03	143	N	-1.0274	0.9643	-0.0148	244.6	-	-	5.7N	113.0W
2045 Aug 27	13:53:26	148	N	1.2061	0.6845	-0.3899	242.4	-	-	8.8S	151.5E
2046 Jan 22	13:01:15	115	P	0.9886	1.0365	0.0550	250.7	51.4	-	20.5N	167.9E
2046 Jul 18	1:04:43	120	P	-0.8692	1.2824	0.2478	299.0	115.3	-	21.8S	14.4W
2047 Jan 12	1:24:51	125	T	0.3317	2.2665	1.2358	338.1	209.6	70.7	22.0N	19.1W
2047 Jul 07	10:34:22	130	Tm	-0.0636	2.7326	1.7529	334.3	219.3	101.6	22.6S	157.3W
2048 Jan 01	6:52:31	135	T	-0.3746	2.2158	1.1297	360.4	215.1	56.7	22.7N	102.4W
2048 Jun 26	2:01:04	140	P	0.6796	1.5841	0.6404	286.5	159.8	-	22.6S	29.6W
2048 Dec 20	6:26:23	145	N	-1.0624	0.9632	-0.1420	282.4	-	-	22.5N	97.3W
2049 May 17	11:25:14	112	N	-1.1337	0.7650	-0.2073	225.0	-	-	20.6S	172.4W
2049 Jun 15	19:12:47	150	N	1.4069	0.2526	-0.6970	132.7	-	-	21.9S	71.9E
2049 Nov 09	15:50:46	117	N	1.1965	0.6821	-0.3541	226.8	-	-	18.2N	118.1E
2050 May 06	22:30:36	122	T	-0.4181	2.1064	1.0779	340.9	206.8	43.8	17.2S	21.4E
2050 Oct 30	3:20:21	127	T	0.4435	2.0356	1.0549	313.9	193.6	35.1	14.2N	54.3W
2051 Apr 26	2:15:02	132	T	0.3371	2.2785	1.2034	365.7	221.7	70.3	13.2S	34.2W
2051 Oct 19	19:10:23	137	T-	-0.2542	2.3719	1.4130	315.0	205.0	84.2	9.9N	68.7E
2052 Apr 14	2:16:40	142	N	1.0629	0.9478	-0.1294	276.8	-	-	8.7S	33.8W
2052 Oct 08	10:44:31	147	P	-0.9727	1.0653	0.0832	257.3	63.9	-	5.3N	164.0W
2053 Mar 04	17:20:42	114	N	-1.0531	0.9334	-0.0796	251.8	-	-	5.1N	102.4E
2053 Aug 29	8:04:22	119	Nx	1.0165	1.0203	-0.0319	278.5	-	-	8.2S	121.1W
2054 Feb 22	6:49:59	124	T	-0.3242	2.2502	1.2780	315.5	201.6	72.7	9.8N	99.2W
2054 Aug 18	9:25:02	129	T	0.2806	2.3817	1.3074	370.4	227.4	83.7	12.7S	140.4W
2055 Feb 11	22:44:49	134	T	0.3526	2.1982	1.2258	313.7	199.1	66.6	14.1N	22.4E
2055 Aug 07	10:51:49	139	P	-0.4769	2.0081	0.9606	347.2	204.2	-	16.8S	161.4W

Section 3: Phases of the Moon: 2031 to 2040

Photo 3-1 Various phases of the Moon over one lunar month. ©2010 F. Espenak

Phases of the Moon

Year	New Moon	First Quarter	Full Moon	Last Quarter
2031			Jan 08 18:26	Jan 16 12:47
	Jan 23 04:31	Jan 30 07:43	Feb 07 12:46	Feb 14 22:50
	Feb 21 15:49	Mar 01 04:02	Mar 09 04:30	Mar 16 06:36
	Mar 23 03:49	Mar 31 00:32	Apr 07 17:21	Apr 14 12:58
	Apr 21 16:57	Apr 29 19:19	May 07 03:40 n	May 13 19:07
	May 21 07:17 A	May 29 11:20	Jun 05 11:58 n	Jun 12 02:20
	Jun 19 22:25	Jun 28 00:19	Jul 04 19:01	Jul 11 11:50
	Jul 19 13:40	Jul 27 10:35	Aug 03 01:45	Aug 10 00:24
	Aug 18 04:32	Aug 25 18:40	Sep 01 09:20	Sep 08 16:14
	Sep 16 18:47	Sep 24 01:20	Sep 30 18:58	Oct 08 10:50
	Oct 16 08:21	Oct 23 07:36	Oct 30 07:33 n	Nov 07 07:02
	Nov 14 21:10 H	Nov 21 14:45	Nov 28 23:18	Dec 07 03:20
	Dec 14 09:06	Dec 21 00:00	Dec 28 17:33	

Year	New Moon	First Quarter	Full Moon	Last Quarter
2032				Jan 05 22:04
	Jan 12 20:07	Jan 19 12:14	Jan 27 12:52	Feb 04 13:49
	Feb 11 06:24	Feb 18 03:29	Feb 26 07:43	Mar 05 01:47
	Mar 11 16:25	Mar 18 20:56	Mar 27 00:46	Apr 03 10:10
	Apr 10 02:39	Apr 17 15:24	Apr 25 15:10 t	May 02 16:02
	May 09 13:36 A	May 17 09:43	May 25 02:37	May 31 20:51
	Jun 08 01:32	Jun 16 03:00	Jun 23 11:32	Jun 30 02:12
	Jul 07 14:41	Jul 15 18:32	Jul 22 18:51	Jul 29 09:25
	Aug 06 05:11	Aug 14 07:51	Aug 21 01:47	Aug 27 19:33
	Sep 04 20:57	Sep 12 18:49	Sep 19 09:30	Sep 26 09:12
	Oct 04 13:26	Oct 12 03:48	Oct 18 18:58 t	Oct 26 02:29
	Nov 03 05:45 P	Nov 10 11:33	Nov 17 06:42	Nov 24 22:48
	Dec 02 20:53	Dec 09 19:09	Dec 16 20:49	Dec 24 20:39

All times are in Universal Time (UT1).
Eclipses at New Moon (solar eclipse) or Full Moon (lunar eclipse), are indicated by these symbols:

Solar Eclipses
T - Total
A - Annular
H - Hybrid
P - Partial

Lunar Eclipses
t - Total (Umbral)
p - Partial (Umbral)
n - Penumbral

Phases of the Moon

Year	New Moon	First Quarter	Full Moon	Last Quarter
2033	Jan 01 10:17	Jan 08 03:34	Jan 15 13:07	Jan 23 17:46
	Jan 30 22:00	Feb 06 13:34	Feb 14 07:04	Feb 22 11:53
	Mar 01 08:23	Mar 08 01:27	Mar 16 01:37	Mar 24 01:50
	Mar 30 17:52 T	Apr 06 15:14	Apr 14 19:17 t	Apr 22 11:42
	Apr 29 02:46	May 06 06:45	May 14 10:43	May 21 18:29
	May 28 11:36	Jun 04 23:39	Jun 12 23:19	Jun 19 23:29
	Jun 26 21:07	Jul 04 17:12	Jul 12 09:29	Jul 19 04:07
	Jul 26 08:12	Aug 03 10:26	Aug 10 18:08	Aug 17 09:43
	Aug 24 21:40	Sep 02 02:24	Sep 09 02:20	Sep 15 17:34
	Sep 23 13:40 P	Oct 01 16:33	Oct 08 10:58 t	Oct 15 04:47
	Oct 23 07:28	Oct 31 04:46	Nov 06 20:32	Nov 13 20:09
	Nov 22 01:39	Nov 29 15:15	Dec 06 07:22	Dec 13 15:28
	Dec 21 18:47	Dec 29 00:20		

Year	New Moon	First Quarter	Full Moon	Last Quarter
2034			Jan 04 19:47	Jan 12 13:17
	Jan 20 10:02	Jan 27 08:32	Feb 03 10:05	Feb 11 11:09
	Feb 18 23:10	Feb 25 16:34	Mar 05 02:10	Mar 13 06:45
	Mar 20 10:14 T	Mar 27 01:18	Apr 03 19:19 n	Apr 11 22:45
	Apr 18 19:26	Apr 25 11:35	May 03 12:16	May 11 10:56
	May 18 03:12	May 24 23:57	Jun 02 03:54	Jun 09 19:44
	Jun 16 10:26	Jun 23 14:35	Jul 01 17:44	Jul 09 01:59
	Jul 15 18:15	Jul 23 07:05	Jul 31 05:54	Aug 07 06:50
	Aug 14 03:53	Aug 22 00:43	Aug 29 16:49	Sep 05 11:41
	Sep 12 16:14 A	Sep 20 18:39	Sep 28 02:57 p	Oct 04 18:05
	Oct 12 07:33	Oct 20 12:03	Oct 27 12:42	Nov 03 03:27
	Nov 11 01:16	Nov 19 04:01	Nov 25 22:32	Dec 02 16:46
	Dec 10 20:14	Dec 18 17:45	Dec 25 08:54	

Year	New Moon	First Quarter	Full Moon	Last Quarter
2035				Jan 01 10:01
	Jan 09 15:03	Jan 17 04:45	Jan 23 20:16	Jan 31 06:02
	Feb 08 08:22	Feb 15 13:17	Feb 22 08:54 n	Mar 02 03:01
	Mar 09 23:09 A	Mar 16 20:15	Mar 23 22:42	Mar 31 23:06
	Apr 08 10:58	Apr 15 02:55	Apr 22 13:21	Apr 30 16:54
	May 07 20:04	May 14 10:28	May 22 04:26	May 30 07:31
	Jun 06 03:21	Jun 12 19:50	Jun 20 19:37	Jun 28 18:43
	Jul 05 09:59	Jul 12 07:33	Jul 20 10:37	Jul 28 02:55
	Aug 03 17:12	Aug 10 21:52	Aug 19 01:00 p	Aug 26 09:08
	Sep 02 01:59 T	Sep 09 14:47	Sep 17 14:23	Sep 24 14:39
	Oct 01 13:07	Oct 09 09:49	Oct 17 02:35	Oct 23 20:57
	Oct 31 02:59	Nov 08 05:50	Nov 15 13:49	Nov 22 05:16
	Nov 29 19:38	Dec 08 01:05	Dec 15 00:33	Dec 21 16:29
	Dec 29 14:31			

All times are in Universal Time (UT1).
Eclipses at New Moon (solar eclipse) or Full Moon (lunar eclipse), are indicated by these symbols:

Solar Eclipses	Lunar Eclipses
T - Total	t - Total (Umbral)
A - Annular	p - Partial (Umbral)
H - Hybrid	n - Penumbral
P - Partial	

Phases of the Moon

Year	New Moon	First Quarter	Full Moon	Last Quarter
2036		Jan 06 17:48	Jan 13 11:16	Jan 20 06:46
	Jan 28 10:17	Feb 05 07:01	Feb 11 22:09 t	Feb 18 23:47
	Feb 27 04:59 P	Mar 05 16:49	Mar 12 09:09	Mar 19 18:39
	Mar 27 20:57	Apr 04 00:03	Apr 10 20:22	Apr 18 14:06
	Apr 26 09:33	May 03 05:54	May 10 08:09	May 18 08:39
	May 25 19:17	Jun 01 11:34	Jun 08 21:02	Jun 17 01:03
	Jun 24 03:09	Jun 30 18:13	Jul 08 11:19	Jul 16 14:39
	Jul 23 10:17 P	Jul 30 02:56	Aug 07 02:49 t	Aug 15 01:36
	Aug 21 17:35 P	Aug 28 14:43	Sep 05 18:45	Sep 13 10:29
	Sep 20 01:51	Sep 27 06:12	Oct 05 10:15	Oct 12 18:09
	Oct 19 11:50	Oct 27 01:14	Nov 04 00:44	Nov 11 01:28
	Nov 18 00:14	Nov 25 22:28	Dec 03 14:08	Dec 10 09:18
	Dec 17 15:34	Dec 25 19:44		

Year	New Moon	First Quarter	Full Moon	Last Quarter
2037			Jan 02 02:35	Jan 08 18:29
	Jan 16 09:34 P	Jan 24 14:55	Jan 31 14:04 t	Feb 07 05:43
	Feb 15 04:54	Feb 23 06:41	Mar 02 00:28	Mar 08 19:25
	Mar 16 23:36	Mar 24 18:39	Mar 31 09:53	Apr 07 11:25
	Apr 15 16:08	Apr 23 03:11	Apr 29 18:54	May 07 04:56
	May 15 05:54	May 22 09:08	May 29 04:24	Jun 05 22:49
	Jun 13 17:10	Jun 20 13:45	Jun 27 15:20	Jul 05 16:00
	Jul 13 02:32 T	Jul 19 18:31	Jul 27 04:15 p	Aug 04 07:51
	Aug 11 10:41	Aug 18 01:00	Aug 25 19:09	Sep 02 22:03
	Sep 09 18:25	Sep 16 10:36	Sep 24 11:32	Oct 02 10:29
	Oct 09 02:34	Oct 16 00:15	Oct 24 04:36	Oct 31 21:06
	Nov 07 12:03	Nov 14 17:59	Nov 22 21:35	Nov 30 06:06
	Dec 06 23:38	Dec 14 14:42	Dec 22 13:38	Dec 29 14:05

Year	New Moon	First Quarter	Full Moon	Last Quarter
2038	Jan 05 13:41 A	Jan 13 12:34	Jan 21 04:00 n	Jan 27 22:00
	Feb 04 05:52	Feb 12 09:30	Feb 19 16:09	Feb 26 06:56
	Mar 05 23:15	Mar 14 03:42	Mar 21 02:09	Mar 27 17:36
	Apr 04 16:43	Apr 12 18:02	Apr 19 10:36	Apr 26 06:15
	May 04 09:19	May 12 04:18	May 18 18:23	May 25 20:43
	Jun 03 00:24	Jun 10 11:11	Jun 17 02:30 n	Jun 24 12:39
	Jul 02 13:32 A	Jul 09 16:00	Jul 16 11:48 n	Jul 24 05:40
	Aug 01 00:40	Aug 07 20:21	Aug 14 22:57	Aug 22 23:12
	Aug 30 10:13	Sep 06 01:51	Sep 13 12:24	Sep 21 16:27
	Sep 28 18:57	Oct 05 09:52	Oct 13 04:22	Oct 21 08:23
	Oct 28 03:53	Nov 03 21:24	Nov 11 22:27	Nov 19 22:10
	Nov 26 13:47	Dec 03 12:46	Dec 11 17:30 n	Dec 19 09:29
	Dec 26 01:02 T			

All times are in Universal Time (UT1).
Eclipses at New Moon (solar eclipse) or Full Moon (lunar eclipse), are indicated by these symbols:

Solar Eclipses	Lunar Eclipses
T - Total	t - Total (Umbral)
A - Annular	p - Partial (Umbral)
H - Hybrid	n - Penumbral
P - Partial	

Phases of the Moon

Year	New Moon	First Quarter	Full Moon	Last Quarter
2039		Jan 02 07:37	Jan 10 11:45	Jan 17 18:42
	Jan 24 13:36	Feb 01 04:45	Feb 09 03:39	Feb 16 02:36
	Feb 23 03:18	Mar 03 02:15	Mar 10 16:35	Mar 17 10:08
	Mar 24 17:59	Apr 01 21:55	Apr 09 02:53	Apr 15 18:07
	Apr 23 09:35	May 01 14:07	May 08 11:20	May 15 03:17
	May 23 01:38	May 31 02:24	Jun 06 18:48 p	Jun 13 14:16
	Jun 21 17:21 A	Jun 29 11:17	Jul 06 02:03	Jul 13 03:38
	Jul 21 07:54	Jul 28 17:50	Aug 04 09:56	Aug 11 19:36
	Aug 19 20:50	Aug 26 23:16	Sep 02 19:23	Sep 10 13:45
	Sep 18 08:23	Sep 25 04:53	Oct 02 07:23	Oct 10 08:59
	Oct 17 19:09	Oct 24 11:50	Oct 31 22:36	Nov 09 03:46
	Nov 16 05:46	Nov 22 21:17	Nov 30 16:49 p	Dec 08 20:44
	Dec 15 16:32 T	Dec 22 10:01	Dec 30 12:37	

Year	New Moon	First Quarter	Full Moon	Last Quarter
2040				Jan 07 11:05
	Jan 14 03:25	Jan 21 02:21	Jan 29 07:54	Feb 05 22:32
	Feb 12 14:24	Feb 19 21:33	Feb 28 00:59	Mar 06 07:19
	Mar 13 01:46	Mar 20 17:59	Mar 28 15:11	Apr 04 14:06
	Apr 11 14:00	Apr 19 13:37	Apr 27 02:38	May 03 20:00
	May 11 03:28 P	May 19 07:00	May 26 11:47 t	Jun 02 02:17
	Jun 09 18:03	Jun 17 21:32	Jun 24 19:19	Jul 01 10:18
	Jul 09 09:15	Jul 17 09:16	Jul 24 02:05	Jul 30 21:06
	Aug 08 00:26	Aug 15 18:36	Aug 22 09:09	Aug 29 11:16
	Sep 06 15:13	Sep 14 02:07	Sep 20 17:43	Sep 28 04:41
	Oct 06 05:26	Oct 13 08:41	Oct 20 04:50	Oct 28 00:27
	Nov 04 18:56 P	Nov 11 15:23	Nov 18 19:06 t	Nov 26 21:07
	Dec 04 07:33	Dec 10 23:30	Dec 18 12:16	Dec 26 17:02

All times are in Universal Time (UT1).
Eclipses at New Moon (solar eclipse) or Full Moon (lunar eclipse), are indicated by these symbols:

Solar Eclipses	Lunar Eclipses
T - Total	t - Total (Umbral)
A - Annular	p - Partial (Umbral)
H - Hybrid	n - Penumbral
P - Partial	

Photo 3–2 The changing phases of the Moon. ©2010 F. Espenak

The phases of the Moon for the entire 21st Century can be found at the AstroPixels.com website:

www.astropixels.com/ephemeris/moon/phases2001gmt.html

Books by the Author

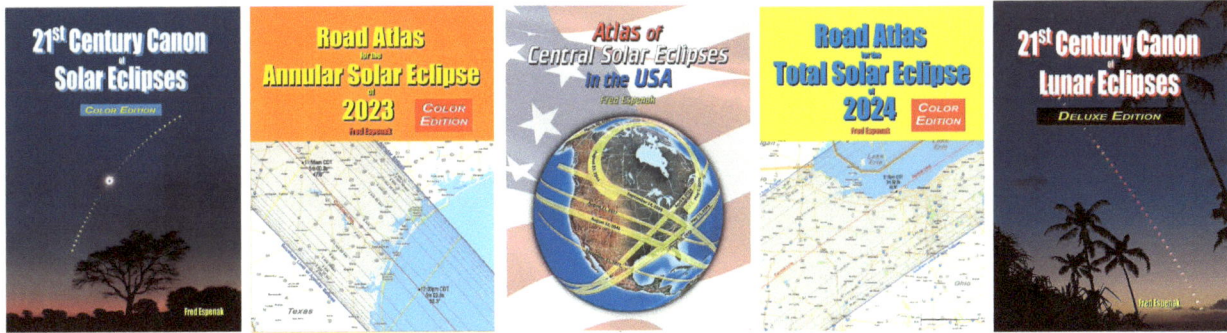

Below is a partial list of some of the books Fred Espenak has written through Astropixels Publishing:

21st Century Canon of Solar Eclipses

The complete guide to every solar eclipse occurring from 2001 tom 2100 (224 eclipses in all). It includes information and maps for all total, annular, hybrid, and partial eclipses. A special world atlas shows detailed full page maps of all central eclipse paths (total, annular and hybrid).

Road Atlas for the Annular Solar Eclipse of 2023

Detailed road maps of the entire eclipse path from the western USA, through Mexico, Central and South America. Information printed on the maps makes it easy to estimate the duration of annularity from any location in the eclipse path. This is the next annular solar eclipse visible from the USA.

Atlas of Central Solar Eclipses in the USA

When was the last total eclipse through the USA and when is the next? How often do they happen? What total eclipse tracks passed across the USA during the 17th, 18th, and 19th centuries, etc., and what states did they include? And how often is a total solar eclipse visible from each of the 50 states? The Atlas of Central Solar Eclipses in the USA answers all of these questions and more with hundreds of maps and tables.

Road Atlas for the Total Solar Eclipse of 2024

This book contains detailed road maps of the entire eclipse path from Mexico, through the USA and Canada. Information printed on the maps makes it easy to estimate the duration of totality from any location in the eclipse path. This is the next total solar eclipse visible from the USA.

21st Century Canon of Lunar Eclipses

The complete guide to every lunar eclipse occurring from 2001 tom 2100 (228 eclipses in all). It includes information and maps for all total, partial, and penumbral eclipses. The predictions use a new model for Earth's elliptical shadows.

For information on these books and more, visit *Astropixels Publishing*:

http://eclipsewise.com/pubs/index.html